SpringerBriefs in Applied Sciences and Technology

Thermal Engineering and Applied Science

Series Editor

Francis A. Kulacki, Minnesota, USA

More information about this series at http://www.springer.com/series/8884

William F. Mohs • Francis A. Kulacki

Heat and Mass Transfer in the Melting of Frost

 Springer

William F. Mohs
Christchurch, New Zealand

Francis A. Kulacki
Department of Mechanical Engineering
University of Minnesota
Minneapolis, MN, USA

ISSN 2191-530X   ISSN 2191-5318   (electronic)
SpringerBriefs in Applied Sciences and Technology
ISBN 978-3-319-20507-6   ISBN 978-3-319-20508-3   (eBook)
DOI 10.1007/978-3-319-20508-3

Library of Congress Control Number: 2015942245

Springer Cham Heidelberg New York Dordrecht London

Printed on acid-free paper

Springer International Publishing AG Switzerland is part of Springer Science+Business Media (www.
springer.com)

Preface

An important problem in the refrigeration industry is the formation and removal of frost layers on subfreezing heat exchanger surfaces of air coolers. Frost is a porous structure of ice and air pockets that grows on the finned surfaces of the heat exchanger. It directly diminishes the performance and efficiency of the entire cooling system by increasing resistance to air flow and heat transfer. To return the system to pre-frosted performance, the frost layer is melted by surface heating. This method is inherently inefficient, with the majority of the applied energy being lost to the surrounding environment.

A validated model for heat and mass transfer in frost melting and defrost efficiency is developed based on local conditions at a vertical surface. Noninvasive measurements of frost thickness are taken from digitally reduced in-plane and normal images of the frost layer through time. The visual method eliminates disturbances caused by contact measurements at the frost–air interface. Thermophysical properties and porosity of the frost are also estimated from digital image analysis. Visual data are combined with temperature and heat flux measurements to produce a detailed analysis of the frost and defrost processes. Heat and mass transfer models are constructed from these measurements, and the defrost process is analyzed through its distinctly defined stages: vapor diffusion, liquid permeation, and dry out.

Frost porosity at the start of defrost has a significant impact on the vapor diffusion stage wherein a large portion of the applied heat to the surface is absorbed by the frost layer. Frost layers with low porosity (high density) absorb more of the applied heat and exhibit higher defrost efficiency. Heat and mass transfer through sublimation during this stage is insignificant.

The second stage of defrost is dominated by melting and permeation of the melt liquid into the overlying frost. Digitally computed frost front velocity is found to vary with the supplied heat transfer rate and frost porosity, which compares well to the visual measurement. Higher heat transfer results in a larger melt velocity and thus shortened defrost time. Lower frost porosity has the effect of increasing the defrost time. As in the first stage of melting, the effects of sublimation are found to be negligible compared to the overall heat and mass transfer. Defrost efficiency for this stage is nearly 100 % with little heat lost to the surroundings.

Evaporation dominates the final stage of defrost. This stage inherently has the low defrost efficiency because most of the supplied heat is lost through sensible heat exchange with the ambient air. Latent heat exchange is correlated to an area reduction of the water droplets, which is expressed by a mass transfer coefficient. A general improvement in the defrost efficiency can be achieved by limiting the duration of the final stage, but this can result in residual moisture on the surface that will be refrozen in the subsequent cooling cycle.

The authors extend their appreciation and thanks to the Ingersoll-Rand/Thermo King Corporation, Bloomington, Minnesota, for the support of this research as the senior author completed his graduate studies at the University of Minnesota.

Christchurch, New Zealand   William F. Mohs
Minneapolis, MN, USA   Francis. A. Kulacki

Contents

Chapter 1
Introduction

Abstract The impact of frost formation on refrigeration performance is described, and several current methods for frost removal are discussed. Defrost efficiency is generally defined for use at either the system or surface levels. An example of the economic impact of defrost is given for U.S. transportation refrigeration units.

Keywords Frost layer • Defrost • Defrost efficiency

Nomenclature

c	Specific heat (J/kg K)
COP	Coefficient of performance (1.1)
E_d	Applied energy of defrost (J)
E_{loss}	Lost energy (J)
m	Mass (kg)
T	Temperature (K)

Greek Letters

η_d	Defrost efficiency
$\eta_{d,s}$	Surface defrost efficiency
λ_{if}	Latent heat of fusion (J/kg)

Subscripts

C	Cold, condensate
f	Frost
H	Hot
loss	Loss
s	Surface

© Springer International Publishing Switzerland 2015   1
W.F. Mohs, F.A. Kulacki, *Heat and Mass Transfer in the Melting of Frost*,
SpringerBriefs in Applied Sciences and Technology,
DOI 10.1007/978-3-319-20508-3_1

1.1   Defrost: An Overview

An important problem in the refrigeration industry is the formation and melting of frost layers on the surfaces of low temperature heat exchanger surfaces, commonly referred to as "coils", in the evaporator of the refrigeration system. Water vapor will condense on the surface of the heat exchanger when its surface temperature is below the local dew point of the air-water vapor mixture in contact with it. When the surface temperature is below the freezing point of water, frost forms either by freezing of the water condensate or via direct sublimation to the surface.

Figure 1.1 shows an idealized frost layer on an extended surface (a fin) anchored to a heat exchanger tube. Frost is a porous structure of ice and air pockets of various sizes and geometries. Experiments show the growth process to occur in a predictable pattern: nucleation, tip growth, densification, and ultimately re-nucleation as the growth process repeats. Frost growth and frost properties have been documented in the recent literature with correlations to calculate frost thickness, density, and conductivity as a function of growth time (Janssen et al. 2012a, b).

The frosting of low-temperature heat exchangers has a significant impact on refrigeration system performance and efficiency. The porous frost layer is an insulating layer on the fin surfaces and thus increases the overall resistance to heat transfer. In addition, as the frost layer grows in depth, the free area for airflow across the frontal area of the heat exchanger is reduced, increasing flow resistance and thus decreasing the air mass flow rate through the coil. Figure 1.2 shows a severely frosted evaporator where the flow area has been completely obstructed by frost. The combined effect of increased thermal resistance and reduced air flow rate causes a drop in the evaporator pressure in a vapor compression refrigeration system as it balances at a new operating point. The drop in evaporator pressure causes a decrease in the surface temperature of the evaporator. For the refrigeration cycle shown in Fig. 1.3, the system operates between temperature reservoirs T_H and T_C in an unfrosted condition. As the frost layer thickens, the evaporator temperature drops to a new reservoir temperature T_C', and the coefficient of performance (COP) of an ideal (Carnot) refrigerator decreases. The Carnot COP is given by (1.1),

Fig. 1.1 Frost on a fin anchored to the surface of a heat exchanger tube

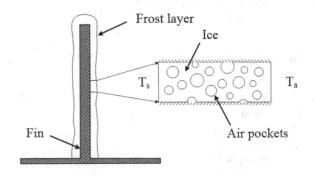

Fig. 1.2 Severely frosted
evaporator (Courtesy of
Ingersoll Rand–Thermo
King Corp.)

Fig. 1.3 Ideal Carnot
refrigerator

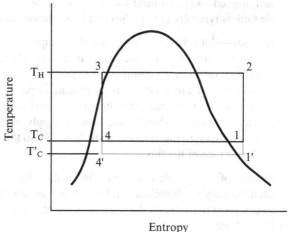

$$COP = \frac{T_C}{T_H - T_C}. \tag{1.1}$$

For example when the high temperature reservoir is at 35 °C and the low temperature reservoir moves from -10 to -15 °C as the coil frosts, the COP shifts from 5.84 to 5.16, a reduction of ~12 % in efficiency. As the system efficiency and performance decrease, it becomes inevitable that the frost layer must be removed by defrosting, i.e., melting, to return refrigeration performance to the pre-frosted condition.

Several methods have been utilized to remove the frost layer during the defrost cycle (Reindl and Jekel 2009; ASHRAE 2014):

Natural Defrost: A common method of defrosting when the conditioned space temperature is near the freezing point of water is to simply turn off flow of coolant, i.e., glycol or refrigerant, and allow the air temperature of the heat exchanger to naturally rise above the melting point. Typically, airflow through the heat exchanger ceases, and the melting process is dominated by free convection, though sometimes air flow is continued to speed up the defrost process at the expense of returning a portion of the moisture to the control space. This method is commonly used for medium temperature display cases in retail applications.

Mechanical Defrost: A mechanical force, such as water or steam jet, is used to remove frost from coils. This method results in fast defrost, and is used in large stationary systems.

Heated Coil Defrost: When the air cooler temperature is well below the freezing point, it is necessary to add heat to melt the frost layer. An external heat source is used to raise the heat exchanger surface temperature above the melting point for a sufficient amount of time to clear it of frost. Water from the melt is collected and drained away. To limit latent and sensible heat gain into the control space, air flow is typically cut off. Several heat sources have been employed:

Hot-gas—Hot refrigerant gas from the compressor of the system is directed to the inlet of the coil. Diverting valves and tubing are required to direct the refrigerant from the compressor to the coil.

Hot-glycol—A hot fluid, such as glycol, is pumped through separate tubing circuits of the exchanger to heat the fin surfaces.

Electrical Heating—Resistance heaters supply heat to the coil. The common methods are to insert heater rods directly into the coil fin pack, or attached to the surface of the fins.

Studies of hot gas defrost have shown that the defrost process goes through predictable stages (Muehlbauer 2006; Donnellan 2007). The average coil temperature is a good indicator of the transition between the stages. The stages of defrost of Fig. 1.4 are,

Pre-defrost: Prior to defrosting, the refrigeration system is operating in the cooling mode but at reduced efficiency. Common indicators used to initiate the defrost cycle are air-side pressure drop across the coil, air-side temperature differential, coil temperature, refrigeration pressure, optical sensor output, and time (a pre-determined operating period).

Stage I—Pre-heat: At the initiation of defrost, liquid refrigerant flow is terminated and replaced by a heat source, e.g., a hot gas refrigerant. Typically the airflow across the evaporator is also stopped to limit latent and sensible heat ingression into the control space. The applied energy is absorbed by the heat exchanger and frost layer, causing the temperatures to rise. Some of the energy is lost through convective heat transfer and conduction to the surrounding support structure of the system. A small portion of the water from the frost surface will be sublimed to the surrounding air.

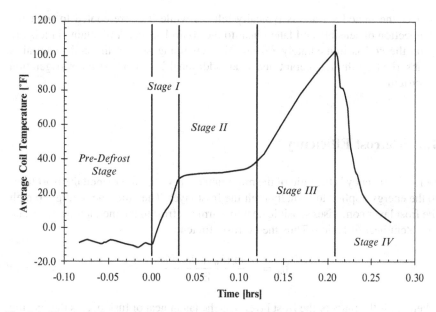

Fig. 1.4 Stages of a hot gas defrost (Courtesy Ingersoll Rand–Thermo King Corp.)

Stage II—Melting and Permeation: Once the interface temperature between the coil
and frost layer rises above the freezing point of ice, the frost layer will begin to
melt. The melt liquid will propagate outward from the surface by permeation.
Initially, the liquid is drawn into the porous frost layer aiding heat transfer. As
the frost layer becomes saturated, some liquid will begin to drain from the coil.
As the interface between the coil surface and frost layer is melted away,
adhesion of the frost layer is greatly reduced and often leads to slumping
(or sloughing) of the frost layer wherein large sections of the frost layer peel
off of the coil and fall into the drain pan. Slumping aids in quick removal of the
frost from the coil, and is more prevalent for frost grown at the tips of the fins.
During melting, the coil surface is characterized by nearly constant temperature
as the majority of the supplied heat is absorbed by the melting frost layer. Some
heat is lost to the surrounding environment through sensible and latent heat
transfer.

Stage III—Dry Out: Once the bulk of the frost layer has become detached from the
coil surface, the coil temperature will begin to rise. Some frost will remain
loosely attached to the coil surface and will continue to melt. In this case, heat
from the fin surface must transfer across an air gap. Frost will continue to slump
from the coil. Retaining liquid water on the coil will be vaporized and lost to the
surrounding air as latent heat.

Stage IV—Re-cool: By some measure (time, coil temperature, optical characteristics, etc.), the evaporator surfaces will be judged to be sufficiently cleared of
frost. Hot-gas heating ends and flow of liquid refrigerant is resumed. Air flow

over the finned surface is typically left off while it is pre-cooled to limit the rejection of sensible and latent heat to the control space. Any retained moisture on the coil is immediately frozen. The remaining heat in the coil material is absorbed by the refrigerant and is an additional heat load on the refrigeration system.

1.2   Defrost Efficiency

Defrost efficiency is the ratio of the minimum energy required to melt the frost layer to the energy applied to actually melt the frost layer. The minimum energy to melt the frost layer comprises sensible heat to warm the frost to the melt temperature and the latent heat of fusion. Thus the defrost efficiency is,

$$\eta_d = = \frac{m_f(\lambda_{if} + c_f\Delta T)}{E_d}, \tag{1.2}$$

where m_f is the mass of the frost layer, λ is the latent heat of fusion, c_f is the specific heat of the frost, ΔT is the difference of the frost temperature and the reference temperature, and E_d, is the energy required for warming and melting the frost layer plus the energy required to warm the surface material, as well as losses to the surrounding ambient environment,

$$E_d = m_f(\lambda_{if} + c_f\Delta T) + m_s c_s\Delta T_s + E_{loss}. \tag{1.3}$$

Combining (1.2) and (1.3),

$$\eta_d = \frac{m_f(\lambda_{if} + c_f\Delta T)}{m_f(\lambda_{if} + c_f\Delta T) + m_s c_s\Delta T_s + E_{loss}}. \tag{1.4}$$

The intent of the defrost process is to return heat exchanger surface to the pre-frosted condition. Thus the sensible heat added to the heat exchanger must be removed through the cooling medium. The energy penalty associated with re-cooling is equal to the heat added to heat the coil material. Thus (1.4) can be modified to,

$$\eta_d = \frac{m_f(\lambda_{if} + c_f\Delta T)}{m_f(\lambda_{if} + c_f\Delta T) + 2m_s c_s\Delta T_s + E_{loss}}. \tag{1.5}$$

From (1.5), it is seen that $\eta_d < 1$ always. The obvious ways to increase defrost efficiency are to limit overheating and heat loss. To evaluate defrost efficiency at the surface, effects of heating and cooling of the surface mass are ignored. The surface defrost efficiency, $\eta_{d,s}$, is attributed only to frost mass and energy lost to the surroundings through heat and mass transfer,

$$\eta_{d,s} = \frac{m_f(\lambda_{if} + c_f \Delta T)}{m_f(\lambda_{if} + c_f \Delta T) + E_{loss}}.$$ (1.6)

The primary disadvantages of all defrost cycles (melting–removal–re-cooling) are that the cooling system is unavailable to maintain the refrigerated space temperature, temperature control is lost, and product temperature may rise above the optimal storage temperature thus reducing shelf life and in extreme cases increase the risk of pathogen growth in perishables. Energy added to the evaporator is parasitic and must be removed from the control space, resulting in longer system run time and more energy consumed.

Defrost cycles are typically chosen in an ad hoc manner. Defrost events are periodically spaced, e.g., every 6 h, and have a set duration. This type of cycle can result in either redundant defrost where minimal frost growth has occurred during the period between defrost cycles or a cycle that is too short to adequately remove the frost altogether prior to termination. To prevent either too few or too short defrosts, it is common to design the defrost algorithm for what is considered the worst-case application, resulting in a defrost cycle that is either too frequent or too long in duration.

As an example, the cost of defrost of a transport refrigeration unit (TRU) is considered. A TRU is used to maintain product temperature as it is shipped from farms to distribution warehouses and on to retail centers, and they tend to experience higher defrost loads compared, say, to stationary refrigeration system owing to larger latent cooling loads. Common sources of the latent cooling load include fresh products, e.g., farm produce, which tend to be loaded when warm and moist and thus naturally respire when latent heat is diffusing from it, frequent door opening for local product distribution, poorly insulated cargo containers with significant leak paths between the conditioned and ambient spaces, and general wear and tear leading to breakdown of door seals.

Table 1.1 summarizes the annual economic cost of defrost for the North American population of semi-trailer refrigeration systems, estimated in 2012 at 333,000 installed and operating units. Ingersoll Rand–Thermo King Corporation conducted a field survey and found that 78 % of the systems operate in the fresh condition, while the remaining 32 % carry frozen loads. Average defrost frequency and defrost duration were found to be dependent on the ambient condition of the refrigerated space. Defrost frequency for fresh conditions is $0.14 \, h^{-1}$ with an average duration of 0.09 h. Defrosting of frozen loads had a greater frequency ($0.24 \, h^{-1}$) and a longer duration (0.18 h). Assuming an average run duration of 2500 h/system-y, a recovery period equal to the defrost duration, an average defrost fuel consumption of 0.6 gal/h, and an average fuel cost of \$3.44 gal^{-1},[1] the total annual cost to defrost the population of systems was \$76.5 million. Assuming defrost efficiency of 10 %

---

[1] Twelve month average price (Nov 2011–Oct 2012) for tax-exempt diesel fuel. Source: U.S. Energy Information Administration, Washington DC, http://www.eia.gov/dnav/pet/pet_pri_gnd_dcus_nus_w.htm.

Table 1.1 Annual economic cost for defrosting North American semi-trailer TRU's[a]

	Frozen	Fresh
Trailer population (N)	106,560	226,440
Average run cycle (h/N)	2500	2500
Defrost/hour (1/h)	0.24	0.14
Average defrost duration (h)	0.18	0.09
Recovery duration (h)	0.18	0.09
Total duration (h)	0.36	0.17
Total defrost hours (h)	23,088,000	13,944,741
Fuel consumption (gal/h)	0.6	
Total defrost fuel (gal)	13,852,800	8,366,845
Fuel cost (US$/gal)	$3.44	
Economic cost (US$/gal)	$47,695,190	$28,807,047
Total economic cost (US$/year)	$76,502,237	
Defrost efficiency (%)	10	80
Minimum defrost cost (US$/year)	$4,769,519	$23,046,637
Recoverable cost (US$/year)	$42,925,671	$5,761,409
Total recoverable cost (US$/year)	$48,687,080	

[a]Data by courtesy of Ingersoll Rand.–Thermo King Corporation

for frozen product and 80 % for fresh (Donnellan 2007), a total of $48.7 million in recoverable fuel costs would have been achievable with a more efficient defrost. The majority of the fuel cost is for the frozen condition where defrost efficiency has been found to be substantially lower than for the fresh condition. It is worth noting that transport refrigeration is a small portion of the refrigeration market, and the total economic cost for the entire refrigeration markets globally would be several orders higher.

References

ASHRAE (2014) ASHRAE Handbook - Refrigeration. Atlanta, GA: American Society of Heating, Refrigeration, and Air Conditioning Engineers, Atlanta

Donnellan W (2007) Investigation and optimization of demand defrost strategies for transport refrigeration systems. Doctoral dissertation, Galway-Mayo Institute of Technology, Galway

Janssen DD, Mohs WF, Kulacki FA (2012a) Modeling frost growth—a physical approach. In: Proceedings of the 2012 ASME summer heat transfer conference, paper no. HT2012-58054

Janssen DD, Mohs WF, Kulacki FA (2012b) High resolution imaging of frost melting. In: Proceedings of the 2012 ASME summer heat transfer conference, paper no. HT2012-58061

Muehlbauer J (2006) Investigation of performance degradation of evaporators for low temperature refrigeration applications. Master's thesis, University of Maryland, College Park

Reindl DT, Jekel TB (2009) Frost on air-cooling evaporators. ASHRAE J 51:27–33

Chapter 2
State-of-the-Art

Abstract The majority of the research conducted over the past five decades has concentrated on the description of frost formation and growth. The prevailing ambient environment greatly influences frost morphology. Several models have been proposed to describe time-variant physical properties and growth of the frost layer, and several researchers have developed frosted fin models to predict the thermal performance of heat exchangers. Experiments have visualized the growth of frost on simple and finned surfaces, as well as, quantified the degradation of the system performance and efficiency under frosted conditions. Recently, studies have been completed to experimentally determine the heat load imposed on the refrigeration system during defrosting and recovery cycles. There have been relatively few models proposed to predict the heat transfer in defrost, with very little analysis of mass transfer. In this chapter we examine several relevant modeling efforts on frost formation and defrost.

Keywords Frost formation • Frost growth • Defrost • System effects of defrost

2.1 Frost Formation and Growth

Iragorry et al. (2004) have compiled an extensive review of the research conducted on modeling frost growth and properties since the earlier review by O'Neal and Tree (1985). The intent of their review is to summarize the modeling work completed on frost formation and growth as it is applied to low temperature evaporator application. Prior studies have shown that frost growth happens in distinct phases: drop-wise condensation (DWC), solidification and tip-growth (STG), densification and bulk growth (DBG). During the initial DWC period, water droplets condense as a sub-cooled liquid on the cold surface. The droplets form into ice particles after a characteristic time has been reached. The characteristic time is a function of ambient conditions (dry bulb temperature and dew point), cold surface temperature, and surface energy. Once the ice droplets form on the surface, the STG stage begins. The effect of the air movement across the surface is most pronounced during this stage. An increase in the bulk stream velocity causes tip growth to slow but does not affect overall frost properties (density, conductivity,

© Springer International Publishing Switzerland 2015
W.F. Mohs, F.A. Kulacki, *Heat and Mass Transfer in the Melting of Frost*,
SpringerBriefs in Applied Sciences and Technology,
DOI 10.1007/978-3-319-20508-3_2

etc.). Once the DBG stage is reached, the frost layer resembles homogeneous porous medium with its properties dependent upon ambient and surface conditions.

Property and parameter models are divided into categories: thermal conductivity, frost density, frost thickness, and heat transfer coefficient. Thermal conductivity is generally correlated as a function of frost density, which is related to the porosity of the frost layer. Theory shows that thermal conductivity is not a function of density alone but also a function tortuosity. Conductivity in low temperature applications is typically ~0.15 to 1.0 W/m K. Density correlations are usually a function of bulk air temperature, the temperature difference between the free air steam and surface, and the humidity ratio of the free stream. Higher defrost density and lower porosity are generally seen for lower bulk temperatures, and density increases with growth time. Frost density is typically in the range of 100–400 kg/m^3 in low temperature applications. Frost thickness, like density, is generally dependent upon the temperature difference and free stream humidity ratio.

Existing correlations for frost growth show rapid thickening of the frost layer during the initial growth period, and slower thickening as the layer becomes developed at later time periods. Heat transfer correlations often take two forms: those based on the energy exchange at the cold surface and those based upon the energy transfer at the frost–air interface. The first includes both the sensible and latent heat transfer effects and results in a higher heat transfer coefficients then for an unfrosted condition. The second reflects only sensible heat convection from the surface, and all models reviewed are from this category. Generally the heat transfer correlations are found to be time invariant, and are based upon the Reynolds number of the flow of the ambient (moist) air over the frost layer. Heat transfer rates on a frosted surface are higher than those on an unfrosted surface, and this increase is attributed to the surface roughness created by the frost.

In addition to parametric models, a number of analytical growth models have been developed. These models take both integral and differential forms, and most of them solve the coupled energy and mass transfer equations to predict the growth of the frost thickness. Mass transfer is modeled as a vapor diffusion process wherein water molecules are directly sublimed to the growth region. They can further be divided into those that assume the solid–vapor interface is at the saturation equilibrium condition defined by the Clausius–Clapeyron equation. Some recent studies (Sherif et al. 2001) have assumed super saturation of the water vapor at the growth interface, but their results do not present strong evidence of improved accuracy compared to models that assume a saturated interface at the growth region. Most of the models presented in the literature are limited to the later tip growth and densification stages of frost growth phase, and require an initial guess of the nucleation growth stage.

The majority of the currently available parametric models have been generated from a limited data set, narrowing the applicability of resulting correlations. Most of them are also limited to conditions where the air temperature is above the freezing point of water and are not considered applicable for low-temperature refrigeration applications. Iragorry et al. (2004) does not indicate which of the analytical models reviewed produces the best results, though an absolute error of about 15 % is common when comparing prediction to measurement. Analytical models are typically limited to the later frost growth stages, and generally require some prior assumption about the nucleation process.

Lee and Ro (2005) propose a model to predict the thickening and densification of the frost layer on a flat plate. It assumes the frost layer is a porous medium of uniform porosity which is supported by observations of previous researchers. The model predicts the diffusion of water vapor into the frost layer and growth of the frost. The model is evaluated for two boundary conditions at the frost–vapor interface. One assumes the frost surface is saturated and the gradient of the vapor pressure within it is at equilibrium with the vapor state. Results show favorable comparison to experimental results (Lee and Ro 2005), but only when the initial porosity is correctly chosen. To remedy the dependence on initial porosity, a modified model is evaluated where water vapor is assumed to be super saturated within the frost layer. Two parameters are reduced from the model: super saturation degree and the diffusion resistance factor. When compared to experimental results, a clear relationship between the degree of saturation and porosity is seen, with a decrease in saturation for higher frost porosity. A correlation between frost porosity and diffusion factor is not seen however.

Kondepudi and O'Neal (1987) review the effects of frost growth on heat exchanger performance. They divide the papers reviewed by four key parameters that effect heat exchanger performance: fin efficiency, overall heat transfer coefficient, pressure drop, and surface roughness. At the time of their review article, limited results were available for frosted heat exchangers owing to the complexity and variety of the geometries commonly found in industry. Of the fin efficiency models they review, they consider the model proposed by Sanders (1974) as the best although it was not validated experimentally. They find a large variability in the overall heat transfer coefficient predicted by the models reviewed. Experimental studies have shown an increase in the overall heat transfer coefficient during the early frost formation stages, followed by a decrease as the frost layer thickens and insulates the surface. Both analytical and empirical models gave no definitive conclusion as to which model most accurately captures the phenomena. The increase in airside pressure drop of the heat exchanger owing to reduction of the flow area caused by the growing frost layer is found to be the most significant parameter affecting heat exchanger performance. The performance reduction is mainly due to the decrease in airflow rate caused by the additional backpressure on the fan. There is limited discussion on the attempt to model the time dependency of frost growth as a function of area reduction, and there is no discussion on using flat surface correlations to build such a model. There is also limited discussion on the effect of a frost layer on surface roughness. The presumption is that the frost layer is randomly distributed on the surface, and this should increase the surface roughness. The increase in the overall heat transfer coefficient has been shown to be the result of this increased roughness, but as the layer thickens it appears to have a minimal effect on the heat exchanger performance.

Based on the gaps found during their review, Kondepudi and O'Neal (1993a, b) completed a study to develop a comprehensive frosted direct expansion heat exchanger model and to validate it against experimental results. The heat exchanger model employs the frost fin efficiency model initially developed by Sanders (1974) for a vertical plate. For the simulation, the heat exchanger is divided into elemental sections, where the frost and refrigerant properties are assumed constant. At each element the frost thickness and associated heat exchanger thermal performance and airside pressure drop are calculated in a quasi-steady state fashion. The temperature is assumed to vary allowing the frost thickness to vary depending upon the fin temperature. A disadvantage of the model is the need to assume an initial frost height and density on the heat exchanger at the start of the simulation, but these assumptions are needed for numerical stability. When compared to the experimental results, the model under predicts the rate of frost accumulation and airside pressure drop, which are directly related as a higher level of frost accumulation (thickness) reduces the air flow and increases the pressure drop. The model predicts the heat transfer rate of the heat exchanger fairly well, generally within 15–20 %. The authors recommend further work to improve the mass transfer of the moisture in the frost layer.

Sommers and Jacobi (2006) analytically solve the problem of a frosted plain fin on a tube. The sector method is used to solve for the frosted fin efficiency in the radial direction within the fin, while the temperature in the frost layer is a function of the radial and axial directions. This assumption is supported by the fact that heat conduction in the fin is substantially greater than in the frost layer, and the thickness of the fin is less than that of the frost layer. Their solution allows for the computation of the fin efficiency and is within 1 % of the exact solution for a non-frosted case. The model has advantages over earlier attempts, namely including two-dimensional heat condition in the frost layer, which allows the use of a modified air side heat transfer coefficient that includes effects of the latent and sensible heat transfer from the air to the frost layer. A disadvantage of the model is the need to assume the frost thickness and conductivity, which are typically not known a priori.

2.2   Defrost

Several fundamental models have been proposed for describing the mechanisms of the defrost process at the surface level. One of the earliest attempts to formulate the analytical solution to the defrost problem is that of Sanders (1974), which stands as the benchmark of the field. Sanders' approach is to formulate the defrost problem as a one-dimensional heat balance at the surface. He considers the case where heat is applied to the surface, solving for the case of hot gas defrost (modeled as a constant surface temperature), as well as an electric defrost (modeled as constant heat flux). The key underlying assumption of Sanders' work is that the melt liquid is drawn into the porous frost layer. The solution of the model allows Sanders to determine the defrost time, heat input, and ultimately defrost efficiency for frost layers of varying thickness.

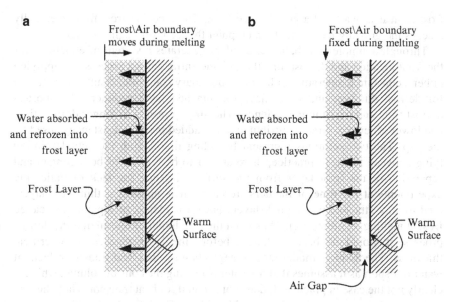

Fig. 2.1 Sanders' models of frost melting (Sanders 1974). (a) Absorption with contact model. (b) Air gap model

Based on the limited qualitative evidence available, Sanders proposed two models to describe the melting process. The first assumes that the frost layer remains attached to the heated surface throughout the defrost process. The melt water is absorbed into the frost layer and re-freezes (Fig. 2.1a). In this model, the frost layer maintains contact with the surface and the exterior air–frost boundary moves towards the surface as the melting process proceeds. The second model (Fig. 2.1b) assumes that melt liquid is drawn into the frost layer leaving behind an air gap at the boundary between the surface and frost. The exterior air–frost boundary stays fixed during the melting process, while the air gap increases as the melting process progresses. Heat is transferred across the gap by natural convection.

For each case, Sanders proposes two phases for the defrost process. The first comprises sensible heat gain of the surface and the frost layer. At the start of the simulation, the wall and frost temperature are assumed to be at an equilibrium temperature below the melt temperature of the frost. When heating is initially applied, most of it is absorbed by the fin material and frost layer, with some lost to loss to the ambient air by convection. When the surface temperature reaches 0 °C, the frost begins to melt, and the second phase of the process begins wherein the melt liquid is drawn into the porous frost layer through capillary action. When the frost layer remains in contact with the surface, heat transfer by conduction is considered, and the air–frost front moves toward it until completely melted. When an air gap is formed, the model assumes that the melt liquid is drawn into the porous frost layer and freezes into ice at the air–frost interface. Thus there is a shrinking frost front near wall and a growing ice front. The receding frost front eventually reaches the growing ice front, at which point the ice layer begins to melt.

Through a simple heat balance, Sanders generates a system of equations for the front location and defrost time through the entire defrost process. By applying either a constant temperature or heat flux boundary condition, results are obtained for defrost duration and efficiency. An attached layer produces the shortest defrost time and maximum efficiency. The air gap results in longer defrost time and lower efficiency directly attributed to the added thermal resistance caused by the gap. The two solutions represent bounding limits, with the actual solution lying between them. In practice, there appear to be regions of both contact and separation of the frost layer from the surface. Despite the lack of validating experiments at the time of Sanders' research, it is noted that thin frost layers tend to stay attached while thick layers have a tendency to separate (so called frost sloughing, or slumping). Thus from the standpoint of optimizing the defrost process, it would be best to defrost before the layer reaches some critical thickness. The Sanders model largely neglects mass and heat transfer effects of water transport as it assumes that no water is leaving the control volume, which is clearly not the case in practice. It does not account for heat transport via either the melt liquid or water vapor that would sublime from the frost surface to the surrounding ambient air. The effect of heat transfer by melt liquid into the frost layer by capillary forces is also neglected.

Aoki et al. (1988) analytically and experimentally investigated the melting process of a porous snow layer that was heat from below. While the thickness of the snow layer is significantly greater than that of a frost layer, a number of significant physical phenomena are similar. First, the melt process of the snow layer proceeds through a number of distinct stages, also seen in the melting of a frost layer. In addition, the factors affecting the permeability of the melt liquid to be drawn into the porous layer are a result of the same capillary forces. The height of the permeability layer is directly related to porosity of the snow layer. Another phenomenon similar to what happens in the melt of a frost layer is the refreezing of the melt liquid into an ice layer, or ice lens. The factors affecting refreezing are the thickness and porosity of the snow layer, heat flux at the heated surface, and surrounding air temperature. Intuitively, if the surrounding air temperature is significantly below the freeze point of water, one would expect a portion of the melt liquid to refreeze into ice. The capillary action of the melt liquid drawn into the porous snow layer increases the rate at which heat is transported from the heated surface but also causes a greater loss of heat to the surrounding air at the air–snow interface. The greater heat transfer rate is due to the larger temperature gradient between the surface and surrounding air.

Krakow et al. (1992a) propose a model to describe the hot-gas defrost process of a refrigeration heat exchanger. From previous experimental work, it is noted that the defrost process has both a spatial and temporal component based on uneven frost thickness and uneven heating during defrost. A finite element model applies mass and energy balances for each stage of defrost. Unknown latent heat effects are combined with sensible heat transfer. Parameters needed to complete the analysis are the retained mass of water on the heat exchanger surfaces and the water vaporization rate, and they are determined experimentally (Krakow et al. 1992b) but are assumed constant in the computations. The model produces reasonable agreement against previous experimental data, even though it largely ignores the effect of water drainage.

Krakow et al. (1993a, b) apply their previously developed heat exchanger model in a full hot-gas defrost cycle simulation. In their first paper (1993a), the major components (heat exchanger, compressor, receiver, and throttling device) of an idealized system are described and modeled. The second paper (1993b) focuses on analysis and validation through measurement. The measurements are limited however, and some parameters are indeterminable, such as saturated refrigerant states, a detailed refrigerant mass inventory, and heat storage in components. The simulation captures the overall trends of hot gas defrost and is useful in seeing relative changes rather than absolute values of pressure and mass flow. The model does not capture higher order dynamics, such as when the reversing valve is initially switched. The dynamics of the model are primarily influenced by the mass and energy storage of the high-side system components.

Sherif and Hertz (1998) developed a model of the defrost process on an electrically heated cylindrical tube. The electrical heater is applied at the surface of the tube, and forms the boundary between the frost layer and refrigerant gas. Heat from the electrical heater is conducted into the frost layer and convected into the refrigerant, both of which must be given as input to the model. Resistance of the heat exchanger material is neglected in the analysis. In this model, the frost melt forms at the heated surface and is immediately drained, the frost layer remains attached to the surface, and the frost front moves towards the heated surface. Employing a quasi-steady-state procedure, the model calculates the transient change in frost thickness and frost surface temperature. The defrost process is assumed to be complete when the frost thickness reaches zero. In general, higher heat fluxes decrease the defrost time. The results of the model are greatly influenced by the selection of the heat flux, which is an input to the model. The model is not compared with experimental results, and the accuracy is not discussed though intuitively it appears to be consistent with observations.

Alebrahim and Sherif (2002) propose and numerically solve a two-dimensional defrost model of frosted circular fins on a tube. As with Sherif's earlier work, the surface is electrically heated to melt the frost. Heat is conducted into the fin and frost, as well as convected to the refrigerant within the tube. The defrost process is divided into two stages. The first stage is the calculation of the steady state temperature distribution within the frost layer. The frost layer is assumed to be uniformly distributed on the heat exchanger surfaces and has homogenous properties. The second stage is the defrost phase where heat is applied at the surface. This stage is further divided into two sub-phases; a pre-melt stage where the fin material and frost layer are heated, and a melt stage where the frost layer begins to melt. The enthalpy formulation of the energy equation is used to model the melting process in the frost layer as it offers the advantage of a continuous model formation in the solid and liquid phase domains of the computation. From the calculated enthalpy at each node, the temperature distribution and melt interface can be determined. The defrost process is assumed to be complete once the frost layer has melted along the entire heat exchanger-frost interface.

The Alebrahim–Sherif model is allows for the determination of the temperature distribution in both the fin and frost in two dimensions. Results show that the melt time can be greatly decreased by increasing the supplied heat flux but that the rate of improvement decreases at extremely high heat fluxes. The total energy input is approximately constant, regardless of supplied heat flux and is dominated by the frost and air temperature. The temperature response and variation in the fin is less pronounced when compared to that in the frost layer, though the fin geometry used was a relatively short thick fin. A longer fin would be expected to have a greater temperature variation. The fin is found to be a significant heat sink due to its relatively large thermal capacitance. The results are not compared to any experimental studies, but are consistent with the findings of previous numerical studies indicating the enthalpy method should be able to be applied to defrost problem in more complex geometries.

Na (2003) expanded on the Sanders air gap model to include the effects of dry out of the heat exchanger. In this instance heating is applied well after the frost has melted to vaporize any remaining moisture of the heat exchanger surfaces, thus resulting in a completely clean heat exchanger for the next cooling period. The model assumes that the mass of the retained moisture is known and at the same temperature as the heat exchanger surface. Water retention is largely dependent on heat exchanger geometry (tube spacing, fin spacing, fin enhancements) and orientation. The addition of the dry out phase greatly increases the calculated defrost time, and lowers the overall defrost efficiency. Na did not have experimental data to compare to his results and was therefore unable to substantiate its accuracy. He recommends that the model be used as a general design tool which calculates the worst case for defrost duration and energy input.

Hoffenbecker et al. (2005) develop a transient model to predict the heat and mass transfer effects associated with hot gas defrosting an industrial air-cooler. The predicted defrost duration compares favorably with experimental results of defrost of an evaporator heat exchanger of an industrial refrigeration system. Model predictions for defrost energy input also compare well to that reported in the literature. One interesting finding of the model is the increase in water vapor transport with a decrease in the inlet refrigerant gas temperature. The increase in water vapor transport is attributed to the prolonged defrost duration, which allows more time for the slower mass diffusion process.

Dopazoa et al. (2010) present a transient simulation for the defrost process on a tube-plate fin heat exchanger. The defrost process is divided into six stages. Energy and mass balances are formulated for a representative tube and fin element for each stage of the defrost process. Their model is capable of varying the inlet refrigeration condition to the heat exchanger and is validated against experimental data and shows excellent agreement in defrost duration. A parametric study shows that defrost time is inversely related to the refrigerant mass flow rate. Increasing the inlet refrigerant temperature decreases defrost duration until it exceeds a critical temperature and then is found to have the opposite effect, resulting in an increase in defrost duration.

2.3 System Effects of Defrost

Al-Mutawa et al. (1998a, b, c, d) and Al-Mutawa and Sherif (1998) and conducted an extensive experimental study to determine the heat load effect of defrosting of a typical low-temperature evaporator on the refrigeration system. The aim of the research was to provide insight into system level effects of defrosting, from which improved guidelines could be developed to accurately determine the defrost load on the refrigeration system. Previous researchers have found that defrost cycles can account for up to 15 % of the total heat load on the refrigeration system. Furthermore current methods of defrost have been found to be inherently inefficient, with typically 15–25 % of the supplied heat leaving with the melt liquid and the remainder as parasitic heat loss.

Mago and Sherif (2002a, b) investigated the effect of defrosting of an industrial fan-heat exchanger with frost grown in a super-saturation condition. Super-saturated frost is formed when the moisture content of the air exceeds the saturation concentration at the prevailing air temperature. The frost is characterized by its snow-like appearance. This type of frost tends to form at fin tips on the air entering side of the heat exchanger and has minimal penetration into the heat exchanger. The test apparatus was constructed such that the heat exchanger inlet and outlet could be dampered shut during the defrost process. The intent of the dampers is to limit sensible and latent heat loss to the test room thereby improving the defrost efficiency. A fully dampered defrost produced 43 % improvement in defrost efficiency relative to the undampered heat exchanger defrost. A partially dampered heat exchanger had an 18 % improvement over the undampered case.

Lohan et al. (2005) studied the defrost process in TRUs. Defrost efficiency in a TRU is largely effected by the frost growth conditions with denser frost resulting in a more efficient defrost process. To facilitate a denser frost growth in frozen conditions, they recommended an adaptive defrost strategy where the time between defrost is a variable dependent upon the refrigerated space condition (temperature and humidity), system operation, and cooling demand.

Several studies have investigated the effect of defrost frequency and duration on both the defrost efficiency, as well as overall system efficiency. Sujau et al. (2006) reported the effect of defrost frequency on the performance of a refrigerated cold store. In general defrost frequency does have an appreciable effect on overall system energy use. A longer interval between defrost results in more frost accumulation on the evaporators, which is removed at a greater relative efficiency during the defrost cycle. However the larger interval has a negative effect on the temperature control of the refrigerated space. From the results, it appears the defrost control sequence was not adjusted to attempt to improve the system efficiency. For the system studied, a defrost interval of 8–12 h appears optimal.

Muehlbauer (2006) measured the performance degradation of a simple refrigeration system with a round-tube, plate fin evaporator through several frosting and defrosting cycles. He reports a drop in system capacity through successive defrost cycles and attributes the loss in performance to moisture retained on the heat exchanger surfaces at the conclusion of a defrost cycle. The retained moisture in the form of droplets freezes on the subsequent cool cycle and acts as nucleation sites during the frost growth period, thus speeding up the frosting process. He confirms that the drop in system performance during the frosting period is attributed to the induced thermal resistance by the frost layer and the additional flow resistance on the airside caused by the reduced flow area. A similar conclusion was experimentally confirmed by Xia et al. (2006) for louvered fins.

2.4 Summary

This literature review has identified a number of studies that have investigated the effects of evaporator frosting on system performance. The common conclusion is that as frost accumulates on the evaporator coil, the free airflow space through the coil decreases thereby reducing the mass flow of air through the coil. The lower airflow rate reduces the cooling capacity of the system. Several models have been proposed to predict the rate of frost accumulation during the growth phase, as well as the thermophysical properties of the frost. Growth models are typically characterized by a relatively quick growth of the frost thickness during the initial stage, followed by a slower densification stage where mass is accumulated internally to the frost layer.

Experimental studies have found that current defrost methods to be inherently inefficient, with up to 75–85 % of the energy required lost to the refrigerated space as a parasitic heat load. Defrost efficiency appears to decrease as the evaporator and ambient air temperatures decrease. This reduction in efficiency can be largely attributed to the additional energy input to heat the coil mass from a lower start temperature. Studies have shown that a drop in system performance after repeated defrost cycles is attributed to the retained moisture on the coil at the conclusion of the defrost event.

Most defrost models use a one-dimensional approach to simplify the complexity of the problem. Early models, neglect the effect of mass transfer, while later models combine the heat and mass transfer effects into lumped terms to simplify the governing equations. Several studies demonstrate how the simplified defrost models can be used in larger system simulations to predict defrost energy input and duration.

References

Alebrahim AM, Sherif SA (2002) Electric defrosting analysis of a finned-tube evaporator coil using the enthalpy method. Proc Inst Mech Eng, Part C: J Mech Eng Sci 216(6):655–673

Al-Mutawa NK, Sherif SA (1998) Determination of coil defrosting loads: part V—analysis of loads (RP-622). ASHRAE Trans 104:344–355

Al-Mutawa NK, Sherif SA, Mathur G (1998a) Determination of coil defrosting loads: part III—testing procedures and data reduction (RP-622). ASHRAE Trans 104:303–312

Al-Mutawa NK, Sherif SA, Mathur GD, West J, Tiedeman JS, Urlaub J (1998b) Determination of coil defrosting loads: part I—experimental facility description (RP-622). ASHRAE Trans 104:268–288

Al-Mutawa NK, Sherif SA, Steadham JM (1998c) Determination of coil defrosting loads: part IV—refrigeration/defrost cycle dynamics (RP-622). ASHRAE Trans 104:313–343

Al-Mutawa NK, Sherif SA, Mathur GD, Steadham JM, West J, Harker RA, Tiedeman JS (1998d) Determination of coil defrosting loads: part II—instrumentation and data acquisition systems (RP-622). ASHRAE Trans 104:289–302

Aoki K, Hattori M, Ujiie T (1988) Snow melting by heating from the bottom surface. JSME Int J 31(2):269–275

Dopazoa JA, Fernandez-Seara J, Uhíaa FJ, Diza R (2010) Modeling and experimental validation of the hot-gas defrost process of an air-cooled evaporator. Int J Refrig 33(4):829–839

Hoffenbecker N, Klein SA, Reindl DT (2005) Hot gas defrost model development and validation. Int J Refrig 28(4):605–615

Iragorry J, Tao Y-X, Jia S (2004) A critical review of properties and models for frost formation analysis. HVAC&R Res 10(4):393–420

Kondepudi SN, O'Neal D (1987) The effects of frost growth on extended surface heat exchanger performance: a review. ASHRAE Trans 93(2):258–274

Kondepudi SN, O'Neal DL (1993a) Performance of finned-tube heat exchangers under frosting conditions: part I—simulation model. Int J Refrig 16:175–180

Kondepudi SN, O'Neal D (1993b) Performance of finned-tube heat exchangers under frosting conditions: part II—comparison of experimental data with model. Int J Refrig 16:181–184

Krakow KI, Yan L, Lin S (1992a) Model of hot-gas defrosting of evaporators—part 1: heat and mass transfer theory. ASHRAE Trans 98(1):451–461

Krakow KI, Yan L, Lin S (1992b) Model of hot-gas defrosting of evaporators—part 2: experimental analysis. ASHRAE Trans 98(1):462–474

Krakow KI, Lin S, Yan L (1993a) An idealized model of reversed-cycle hot gas defrosting—part 1: theory. ASHRAE Trans 99(1):317–328

Krakow KI, Lin S, Yan L (1993b) An idealized model of reversed-cycle hot gas defrosting—part 2: experimental analysis and validation. ASHRAE Trans 99(2):329–338

Lee YB, Ro ST (2005) Analysis of the frost growth on a flat plate by simple models of saturation and super saturation. Exp Therm Fluid Sci 29(6):685–696

Lohan J, Donnellan W, Gleeson K (2005) Development of efficient defrosting strategies for refrigerated transportation systems: part I—experimental test facility. In: Proceedings of the international IIR conference on latest developments in refrigerated storage, transportation and display of food products, Amman

Mago PJ, Sherif SA (2002a) Modeling the cooling process path of a dehumidifying coil under frosting conditions. J Heat Transfer 124:1182–1191

Mago PJ, Sherif SA (2002b) Dynamics of coil defrosting in supersaturated freezer air. Proc Inst Mech Eng: Part C, J Mech Eng Sci 212:949–958

Muehlbauer J (2006) Investigation of performance degradation of evaporators for low temperature refrigeration applications. Master's thesis, University of Maryland, College Park

Na B (2003) Analysis of frost formation in an evaporator. Doctoral dissertation, Pennsylvania State University, University Park

O'Neal DL, Tree DR (1985) A review of frost formation in simple geometries. ASHRAE Trans 91(2A):267–281

Sanders CT (1974) The influence of frost formation and defrosting on the performance of air coolers. Dissertation WTHD 63, Delft University of Technology, Delft

Sherif SA, Hertz MG (1998) A semi-empirical model for electronic defrosting of a cylindrical coil cooler. Int J Energy Res 22(1):85–92

Sherif SA, Mago PJ, Al-Mutawa NK, Theen RS, Bilen K (2001) Psychometrics in the supersaturated frost zone. ASHRAE Trans 107(2):753–767

Sommers AD, Jacobi AM (2006) An exact solution to steady heat conduction in a two-dimensional annulus on a one-dimensional fin: application to frosted heat exchangers with round tubes. J Heat Transfer 128:397–404

Sujau M, Bronlund JE, Merts I, Cleland DJ (2006) Effect of defrost frequency on defrost efficiency, defrost heat load, and coolstore performance. Refrig Sci Tech Ser 1:525–532

Xia Y, Zhong Y, Hrnjak PS, Jacobi AM (2006) Frost; defrost; and refrost and its impact on the air-side thermal-hydraulic performance of louvered-fin; flat-tube heat exchangers. Int J Refrig 29:1066–1079

Chapter 3
Multi-stage Defrost Model

Abstract A comprehensive one-dimensional model is developed for heat and mass transfer in each stage of the defrost process. The model describes sublimation, vapor transport, liquid melting and evaporation on a heated vertical surface in each of three distinct stages of the defrost process: diffusion, melting-permeation and dry out. Dimensionless forms of the governing differential equations yield the Lewis, Stefan, and Biot numbers as the key parameters. These parameters for each stage of defrost are slightly different owning to the difference in the dominate heat and mass transfer mechanisms. By analyzing the magnitude of the dimensionless groups, it is possible to determine the relative weight of each term in the governing equation. From such an analysis, the effect of mass transfer due to sublimation in the first and second stages of the defrost process can be neglected. The effects of several limiting cases on the governing equations are evaluated, with simplified equation sets developed for them.

Keywords Defrost • Heat transfer • Mass transfer • Sublimation • Vapor transport • Permeation • Dry out

Nomenclature

a	Conductivity weighting factor
A	Area (m^2)
b_1, b_2	Coefficients in vapor pressure-versus-temperature relation (3.48)
Bi	Biot number, $\delta h/k$
c	Specific heat (J/kg K)
c_p	Specific heat at constant pressure (J/kg K)
D	Mass diffusion coefficient (m/s)
g	Gravitational acceleration (m/s^2)
h	Heat transfer coefficient (W/m^2 K)
h_m	Mass transfer coefficient (m/s)
i	Enthalpy (J/kg)
k	Thermal conductivity (W/m K)

© Springer International Publishing Switzerland 2015

W.F. Mohs, F.A. Kulacki, *Heat and Mass Transfer in the Melting of Frost*,
SpringerBriefs in Applied Sciences and Technology,
DOI 10.1007/978-3-319-20508-3_3

21

Le   Lewis number (α/D)
m   Mass (kg)
m''   Mass flux (kg/m^2 s)
M   Dimensionless mass flux
p   Pressure (Pa)
q''   Heat flux (W/m^2)
R   Ideal gas constant (Pa/mol K)
S   Water content ratio: fraction of frost pore volume containing water (–)
S_c Limited water content (–)
St   Stephan number, $c\Delta T/\lambda$
t   Time (s)
T   Temperature (K)
u   Convective velocity (m/s)
u_0 Initial melt velocity (3.16)
U   Dimensionless velocity (3.56)
w   Fin half thickness (m), Fig. 3.1
W   Width of surface (m), Fig. 3.1
y   Distance from surface (m)

Greek Letters

α Thermal diffusivity, k/ρc (m^2/s)
δ Thickness (m)
ε Porosity (–)
ϕ Dimensionless density (3.37b)
Γ Reciprocal of the Stephan-Lewis number product, 1/StLe
κ Dimensionless thermal conductivity (3.37c)
η Dimensionless distance (3.37b)
π Dimensionless diffusion coefficient (3.37c)
ρ Density (kg/m^3)
μ Viscosity (N s/m)
θ Dimensionless temperature
λ_{if} Latent heat of fusion (J/kg)
$\lambda_{\phi\gamma}$ Latent heat of vaporization (J/kg)
λ_{ig} Latent heat of sublimation (J/kg)
$\tilde{\tau}$ Tortuosity (–)
τ Dimensionless time (–)

Subscripts

0   Initial time
1,2,3   Denoting stage of defrost

a   Air
eff   Effective
f   Frost
fs   Frost surface
i   Ice
lt   Latent
m   Melt
p   Permeation
s   Surface
sn   Sensible
t   Total
v   Vapor
w   Water

Other Symbols

$\perp$ Perpendicular
$\parallel$ Parallel

3.1 Vertical Surface Geometry

It is apparent from the current literature that there is a need to advance the capabilities of accurately modeling the defrost process. At best, the current state of modeling can be used only as a general predictive tool for worst case scenarios and lacks the necessary accuracy to describe the defrost process. Generally current modeling is limited to one-dimensional heat transfer with simplified frost properties and neglects effects of latent heat through either melt drainage or sublimation. Gravitational effects on melt drainage and frost slumping are ignored as well. Current models are limited to a specific stage of the defrost process, and no current model is capable of predicting heat and mass transfer through the entire defrost process.

The present defrost model is specialized to a vertical surface. This geometry permits the formation of a tractable mathematical problem and forms a basic step toward modeling geometries that occur in application, e.g., a multi-fin evaporator. Figure 3.1 shows a plain half fin, which represents a single fin of a common evaporator. The fin is attached to a continuous base where heating is applied. The fin has a length L and a half-thickness w. A frost layer of thickness δ is bonded to the fin surface. The frost is constructed of interlacing ice crystals with internal air pockets and has a bulk porosity ε. There is some debate within the literature whether porosity varies within the frost layer, and our model is constructed such that effects of either a constant or variable porosity can be evaluated. Initially the surface and surrounding air temperature are assumed to be below the melt

Fig. 3.1 Frosted vertical
surface, e.g., a half fin

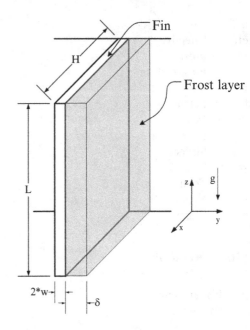

temperature. The temperature within the ice is assumed to vary linearly between the wall and frost surface temperature. The water vapor within air pockets in the frost is assumed to be at the saturation pressure of the local frost temperature. The thin fin approximation is used to model the fin, with only temperature variations along the height of the fin assumed. Heat and mass transfer within the frost layer is assumed to occur predominately in the y-direction. The conductivity of the metal fin is several orders of magnitude greater than that of the frost, and thus temperature gradients in the x-direction are assumed small in comparison to those in the frost layer.

3.2 Stage I Defrost: Diffusion

When heat is applied to the frosted surface, the increase of surface temperature causes a change in the internal temperature distribution of the frost layer. The temperature change also causes a change in the local vapor pressure, and the small pressure difference will cause water molecules to sublimate. As the local vapor pressure increases above the prevailing bulk vapor pressure, water molecules will diffuse from the ice surface through the frost layer and escape into the surrounding air. Mass transport of water vapor carries latent heat from the frost layer, and sensible heat is lost to the ambient air through convection. One would not expect the overall height of the frost layer to dramatically change during this stage but

Fig. 3.2 Defrost Stage I.
Heat and vapor diffusion

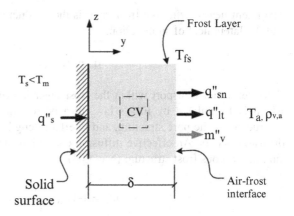

instead see a slight decrease in the density of the frost layer. Figure 3.2 depicts the frost layer with the several flux quantities for heat and mass transfer.

The energy balance over the control volume of Fig. 3.2 is,

$$(\rho c)_f \frac{\partial T}{\partial t} = \frac{\partial}{\partial y}\left(k_f \frac{\partial T}{\partial y}\right) - \frac{\partial q_v''}{\partial y}, \tag{3.1}$$

where the l.h.s. is the rate of change in internal energy, and the r.h.s. expresses energy transport by conduction and sublimation of the frost. The volumetric heat capacity, $(\rho c)_f$, of the frost layer is determined by the arithmetic mean (volume) weighed, porosity, ε,

$$(\rho c)_f = \varepsilon(\rho c_p)_a + (1 - \varepsilon)(\rho c)_i. \tag{3.2}$$

A common approximation for the thermal capacity of frost is to ignore the capacitance of the air, and $(\rho c)_f \approx (1 - \varepsilon)(\rho c)_i$.

Several models have been proposed to determine the effective conductivity of frost. Generally these models relate the effective conductivity to bulk porosity, e.g., Auracher's (1987) empirical correlation using a parallel and perpendicular plate frost structure,

$$\frac{1}{k_f} = \frac{a}{k_{f,\perp}} + \frac{1-a}{k_{f,\|}}, \quad a = 0.42(0.1 + 0.955\rho_f), \tag{3.3a}$$

$$k_{f,\perp} = \left[\frac{\varepsilon}{k_a} + \frac{1-\varepsilon}{k_i}\right], \quad k_{f,\|} = \varepsilon k_a + (1 - \varepsilon)k_i. \tag{3.3b}$$

The latent heat leaving the frost layer is the product of the water vapor mass flux and the latent heat of sublimation,

$$q_v'' = \lambda_{ig} m_v''. \tag{3.4}$$

Water vapor transport within the frost layer is driven by molecular diffusion, which is formulated by Fick's law. In a porous medium, the diffusion path is obstructed by the pore structure, and in effect, lengthens the path along which the diffusion occurs. An effective diffusion coefficient, $D_{v,eff}$, can describe the diffusion in a porous frost structure (Na 2003),

$$m_v'' = -D_{v,eff} \frac{\partial \rho_v}{\partial y}. \tag{3.5}$$

Generally it is assumed that the water vapor within the pores is saturated at the local frost interface temperature. Applying the ideal gas law, (3.5) becomes,

$$m_v'' = -D_{v,eff} \frac{\partial}{\partial y} \left(\frac{p_v}{RT} \right). \tag{3.6}$$

With local thermal equilibrium (Wiederhold 1997), the vapor pressure and can be determined by,

$$p_v = 6.1115 \left(1.003 + 4.18 \times 10^{-6} p_t \right) \exp \left[22.452 T / 272.55+T \right] \tag{3.7}$$

The effective diffusion coefficient is related to the bulk diffusion coefficient by the bulk porosity and tortuosity, $D_{v,eff}/\varepsilon D_v$, which describes the path through the porous structure. Lee and Ro (2005) recommend the following for the diffusion coefficient and tortuosity,

$$D_v = 9.26 \times 10^{-7} \frac{1}{p_a} \left(\frac{T^{2.5}}{T + 245} \right), \tag{3.8}$$

$$\tilde{\tau} = \frac{\varepsilon}{1 - (1 - \varepsilon)^{0.5}}. \tag{3.9}$$

Equations (3.1), (3.4), and (3.6) result in the following differential equation for temperature,

$$\left(\rho c_p \right)_f \frac{\partial T}{\partial t} = \frac{\partial}{\partial y} \left(k_f \frac{\partial T}{\partial y} \right) + \frac{\partial}{\partial y} \left(\lambda_{ig} D_{v,eff} \frac{\partial \rho_v}{\partial y} \right), \tag{3.10}$$

with boundary conditions,

$$q_s'' = -k_f\frac{\partial T}{\partial y}\bigg|_{y=0}, \tag{3.11a}$$

$$k_f\frac{\partial T}{\partial y}\bigg|_{y=\delta} = h(T_a - T_{fs}) + \lambda_{ig}h_m\left(\rho_{v,a} - \rho_{v,fs}\right), \tag{3.11b}$$

where q_s'' is the supplied heat flux at the surface, and T_a, and T_{fs} are the ambient air and frost surface temperatures respectively. The initial temperature distribution, T (y,0), is determined through the solution of the steady state heat conduction equation for a given heat flux and surface temperature. The convective heat transfer coefficient, h, and the mass transfer coefficient, h_m, can be estimated via the heat and mass transfer analogy and are related by the Lewis number, Le_a, and the specific heat of air (Hao et al. 2005),

$$h_m = \frac{h}{\rho_a c_{p,a}Le_a^{2/3}}. \tag{3.12}$$

The change in density due to the sublimation can be expressed via conservation of mass,

$$\frac{\partial \rho_f}{\partial t} = -\frac{\partial m_v''}{\partial y} \tag{3.13}$$

The final form of the mass balance with two boundary conditions and the initial condition,

$$\frac{\partial \rho_f}{\partial t} = -\frac{\partial}{\partial y}\left(D_{v,eff}\frac{\partial \rho_v}{\partial y}\right), \tag{3.14a}$$

$$\frac{\partial \rho_v}{\partial t} = 0, \quad y = 0, \tag{3.14b}$$

$$m_v'' = h_m\left(\rho_{v,a} - \rho_{v,fs}\right), \quad y = \delta, \tag{3.14c}$$

$$\rho_f = \varepsilon\rho_a + (1 - \varepsilon)\rho_i, \quad t = 0, \tag{3.14d}$$

where typically the frost density is assumed to be constant across the thickness of the frost layer.

3.3 Stage II Defrost: Melting–Permeation

The frost begins to melt when the surface temperature reaches the melt temperature of the ice. As liquid water accumulates, capillary forces draw the liquid into the small air pockets near the surface creating a layer of water and ice, the permeation

Fig. 3.3 Defrost Stage II.
Melting with liquid
permeation

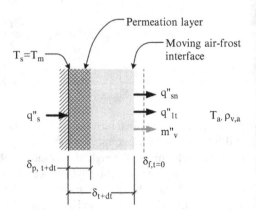

layer (Fig. 3.3). Heat transfer is by movement of the melt liquid and conduction through the ice. Heat transfer will reach a practical limit however because the applied heat flux is finite, and thermal energy will be absorbed by the melting frost at the melt temperature within the permeation layer. As the frost melts and the liquid is drawn into it, the thickness of the frost layer will decrease. Within the permeation layer, water content varies from almost completely liquid near the surface to no liquid in the un-melted portion of the frost layer. Thus there are two moving fronts: a liquid permeation front moving away from the surface and the frost-air front moving toward the surface. A small portion of thermal energy will escape the frost layer through both sensible and latent heat transfer.

Within the frost layer the equation for conservation of energy is similar to that for Stage I with the addition of convective transport to account for heat transfer by the bulk movement of the frost layer. Under the assumption that two-dimensional liquid flow can be neglected, $T = T(y,t)$, and,

$$(\rho c)_f \frac{\partial T}{\partial t} = \frac{\partial}{\partial y}\left(k_f \frac{\partial T}{\partial y}\right) + \frac{\partial}{\partial y}\left(\lambda_{ig} D_{v,\text{eff}} \frac{\partial \rho_v}{\partial y}\right) + (\rho c)_f u \frac{\partial T}{\partial y}. \tag{3.15}$$

The convective velocity, u, arises owing to the reduction in volume due to the melting of the frost layer and is given by,

$$u = \frac{\partial \delta}{\partial t} = -\frac{q_s''}{\lambda_{if}\rho_f}. \tag{3.16}$$

The remaining terms in (3.15) are evaluated as previously described. The mass balance is given by (3.13). The boundary and initial conditions are,

$$T = T_m, \quad m_v'' = 0, \quad y = \delta_p, \tag{3.17a}$$

$$k_f \frac{\partial T}{\partial y}\bigg|_{y=\delta} = h(T_a - T_{fs}) + \lambda_{ig} h_m (\rho_{v,a} - \rho_{v,fs}), \quad y = \delta \qquad (3.17b)$$

$$m_v^{''} = h_m (\rho_{v,a} - \rho_{v,fs}), \quad y = \delta \qquad (3.17c)$$

$$\frac{\partial \rho_f}{\partial y} = \frac{\partial \varepsilon}{\partial y}\bigg|_{t=0}, \quad t = 0, \qquad (3.17d)$$

where the initial frost density profile is defined by the porosity at the conclusion of Stage I.

Aoki et al. (1988) developed a model to describe the penetration of the melt liquid into a porous layer in terms of liquid water content ratio, S. The water content ratio describes the volume fraction of liquid water in the open pockets within the frost layer. Near the heated surface, $S \approx 1$ and decays away from the surface. The time dependence of water content in the permeation layer is,

$$\frac{\partial S}{\partial t} = -\frac{1}{\varepsilon \rho_w} \frac{\partial m_{w,p}^{''}}{\partial y} - u \frac{\partial S}{\partial y}. \qquad (3.18)$$

The r.h.s. describes the contribution of mass flux due to water permeation coming from the surface and the bulk movement of the frost layer towards the surface. The water permeation mass flux, $m_{w,p}^{''}$, is dependent on the frost structure and capillary and gravitational forces. The rate of change in the location of the water permeation front, δ_p, is equal to the rate that water permeates into the frost, the rate at which the water refreezes, and the rate at which the frost is moving toward the surface,

$$\frac{\partial \delta_p}{\partial t} = \frac{m_{w,p}^{''}\big|_{y=\delta_p}}{\varepsilon \rho_w S_c} + \frac{1}{\varepsilon \rho_w S_c} \frac{k_p}{\lambda_{if}} \frac{\partial T}{\partial y}\bigg|_{y=\delta_p} + u, \qquad (3.19)$$

where S_c is the limited water content, the minimum water concentration. Aoki recommends that $S_c \approx 0.1$. The change in density in the permeation layer due to refreezing can be represented by,

$$\delta_p \frac{\partial \rho_p}{\partial t} = -\frac{k_p}{\lambda_{if}} \frac{\partial T}{\partial y}\bigg|_{y=\delta_p}. \qquad (3.20)$$

As ice crystals near wall are melted and the permeation layer becomes fully saturated, a liquid layer forms on the surface. Some of the liquid will drain from the surface owing to gravity (Fig. 3.4). The mass flux of the melt liquid leaving the surface, $m_{w,s}^{''}$, is equal to the difference between the mass flux of the melt liquid generated at the surface, $m_{w,t}^{''}$, and the water permeation mass flux,

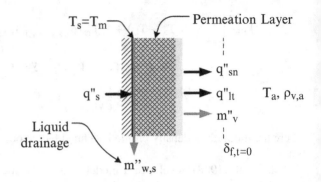

Fig. 3.4 Defrost Stage II. Liquid permeation and drainage

$$m_{w,s}^{''} = m_{w,t}^{''} - m_{w,p}^{''}. \tag{3.21}$$

The total mass flux at the surface is the sum of liquid generated and the mass flux of liquid carried with the movement of the frost layer toward the surface,

$$m_{w,t}^{''} = -u\left(\rho_p + \varepsilon\rho_w S_{y=0}\right). \tag{3.22}$$

Combining (3.20) with (3.21) and (3.15), the melt drainage mass flux is,

$$m_{ws}^{''} = \frac{q_s^{''}}{\lambda_{if}\rho_f}\left(\rho_p + \varepsilon\rho_w S_{y=0}\right) - m_{w,p}^{''}. \tag{3.23}$$

Boundary conditions for the permeation layer are,

$$q_s^{''} = -k_w\frac{\partial T}{\partial y}\bigg|_{y=0}, \quad y = 0, \tag{3.24a}$$

$$T = T_m, \quad S = S_c, \quad y = \delta_p. \tag{3.24b}$$

The initial temperature and vapor distribution are inherited from the end of Stage I. As recommended by Aoki et al. (1988), the initial water content ratio at the surface is taken as $S_c \approx 1$, and $S_c = 0.1$.

Melting continues until either the entire frost layer has melted or the frost pulls away from the surface in a bulk movement called sloughing, or slumping. The tendency for sloughing is dependent upon adhesion forces governed by the surface tension at the frost-solid interface. One would expect the likelihood of sloughing to be greater for thick, dense frost wherein the weight of the frost layer is greater than the adhesion force to the surface. We consider only frost layers that stay in contact with the surface in this book.

3.4   Stage III Defrost: Dry Out

After the frost layer has melted a small portion of the melt liquid will adhere to the surface owing to surface tension. The mass of retained liquid will be a function of inclination angle and surface wettability. Wettability is determined by a force balance between adhesive and cohesive forces of a water droplet. Surfaces that have a high degree of wettability are termed hydrophilic surfaces and tend to form a thin layer of water. Hydrophobic surfaces are low wetting surfaces and tend to form individual droplets of water on the surface. Liquid water will continue to shed from the surface as a falling film, while heating of the surface will cause some of the water to be evaporated. The following analysis is a simplified approach to the model presented by Yan and Lin (1990) for coupled heat and mass transfer of falling film evaporation.

From a mass balance on of the control volume in Fig. 3.5, the rate of change of mass per unit area is equal to the sum of mass flux of the melt liquid falling down the surface and the mass flux by evaporation,

$$\frac{\partial}{\partial t}\left(\frac{m}{A}\right) = m_{w,s}'' + m_v''. \tag{3.25}$$

The mass flux leaving the surface is a simple evaporation process. Assuming the liquid layer is sufficiently thin, temperature variations within the film can be neglected, and in addition, the film temperature is the same as the wall temperature. The mass flux can then be determined by,

$$m_v'' = h_m\left(\rho_{v,a} - \rho_{v,w}\right), \tag{3.26}$$

where h_m is the mass transfer coefficient. The mass flux of the liquid stream can expressed as the product of the film velocity, u_w, and liquid density, ρ_w,

$$m_{w,s}'' = u_w\rho_w. \tag{3.27}$$

Fig. 3.5 Defrost Stage III. Dry out

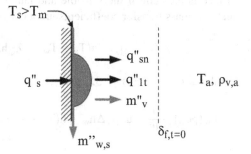

Assuming laminar film flow and negligible inertia effects, the film velocity can be determined by the momentum equation,

$$\frac{\partial}{\partial y}\left(\mu_w \frac{\partial u_w}{\partial y}\right) + \rho_w g = 0 \tag{3.28}$$

The motion of the water film is dependent on the balance of viscous and gravitational forces. Assuming a no slip condition at the wall and small velocity at the melt-air interface, the mean velocity can be determined by integrating (3.28) to obtain,

$$\bar{u}_w = -\frac{\rho_w g}{12\mu_w}\delta_w^2, \tag{3.29}$$

where δ_w is the thickness of the water stream. Combining (3.25)–(3.28), the change in film thickness is,

$$\frac{\partial \delta_w}{\partial t} = -\frac{\rho_w g}{12\mu_w}\delta_w^2 + \frac{h_m}{\rho_w}(\rho_{v,a} - \rho_{v,w}). \tag{3.30}$$

With the assumption that the temperature variation across the liquid film is negligible, a lumped analysis can be applied to the control volume. The rate of temperature change of the film is due to thermal energy leaving with the liquid drainage and the sensible and latent heat exchanges to the ambient air,

$$\delta_w\left(\rho c_p\right)_w \frac{dT}{dt} = q_{w,s}'' + q_v'' + q_s''. \tag{3.31}$$

Energy leaving with the liquid steam, q_s'', can be expressed,

$$q_{w,s}'' = m_{w,s}'' \Delta i_w = \bar{u}_w \rho_w \Delta i_w, \tag{3.32}$$

where Δi_w is the change in enthalpy of the water at the film temperature, and $\bar{u}_w$ is the mean water stream velocity. Similar to earlier analysis, the energy leaving the surface is the sum of the sensible and latent heat transfers, where the convective heat and mass transfer coefficients are analogous via the Lewis number,

$$q_v'' = h(T_a - T_w) + \lambda_{fg}h_m(\rho_{v,a} - \rho_{v,w}). \tag{3.33}$$

Combination of (3.30)–(3.32) gives,

$$\delta_w\left(\rho c_p\right)_w \frac{dT}{dt} = \bar{u}_w \rho_w \Delta h_w + h(T_a - T_w) + \lambda_{fg}h_m(\rho_{v,a} - \rho_{v,w}) + q_s''. \tag{3.34}$$

The initial conditions are the surface temperature and film thickness. The initial film thickness can be estimated from the retained liquid mass.[1] El Sherbini and Jacobi (2006) have developed a detailed retention model where contact angle is used to calculate the volume of an individual droplet. Employing a size-distribution function they determine the retained mass for the entire heat exchanger. Applying the model to a plane fin heat exchanger, they find retained mass of ~120 g/m², which compares favorably with their measurements.

The most notable assumption for Stage III defrost is the neglect of the effect of surface tension on the water hold up phenomena. Equation (3.28) should include a term to describe the resistive force caused by surface tension effects. As mentioned earlier surface tension will cause adhesion to the surface that impedes water drainage. Also the current model does not capture the dynamics of bulk frost movement. A sloughing model will be closely tied to surface tension, as it would be the primary adhesive force acting of the frost.

3.5 Scale Analysis

Each stage of our defrost model can be solved separately. The final state of each stage provides initial condition for the subsequent stage. To gain insight into the relative magnitude of each term in the governing equations, the governing equations are now cast in dimensionless variables. By applying the appropriate dimensionless parameters to the physical dimensions of the problem, the governing equations can be reduced to a simpler form for solution.

Stage I Defrost—Vapor Diffusion: The first stage of the defrost process is dominated by the sensible heating of the frost layer, and sublimation into the ambient air. Recalling the energy equation for Stage I,

$$(\rho c)_f \frac{\partial T}{\partial t} = \frac{\partial}{\partial y}\left(k_f \frac{\partial T}{\partial y}\right) + \frac{\partial}{\partial y}\left(\lambda_{ig} D_{v,eff} \frac{\partial \rho_v}{\partial y}\right). \tag{3.35}$$

Applying the chain rule to the r.h.s. to expand the terms,

$$(\rho c)_f \frac{\partial T}{\partial t} = k_f \frac{\partial^2 T}{\partial y^2} + \frac{\partial k_f}{\partial y}\frac{\partial T}{\partial y} + \lambda_{ig} D_{v,eff} \frac{\partial^2 \rho_v}{\partial y^2} + \lambda_{ig}\frac{\partial D_{v,eff}}{\partial y}\frac{\partial \rho_v}{\partial y}. \tag{3.36}$$

Next we, define the following dimensionless parameters,

---

[1] Factors affecting retention are surface wettability, due to coating and surface finish, obstruction, e.g., louvers and other fin geometries, and fin spacing which can lead to bridging.

$$\theta = \frac{T - T_s}{T_m - T_s} = \frac{T - T_s}{\Delta T}, \quad \tau = \frac{\alpha_f t}{\delta_f^2}, \tag{3.37a}$$

$$\phi = \frac{\rho_v - \rho_{v,s}}{\rho_{v,m} - \rho_{v,s}} = \frac{\rho_v - \rho_{v,s}}{\Delta \rho_v}, \quad \eta = \frac{y}{\delta_f}, \tag{3.37b}$$

$$\pi = \frac{D_{v,eff} - D_{v,eff,s}}{D_{v,eff,fs} - D_{v,eff,s}} = \frac{D_{v,eff} - D_{v,eff,s}}{\Delta D_{v,eff}}, \quad \kappa = \frac{k_f - k_{f,s}}{k_{f,fs} - k_{f,s}} = \frac{k_f - k_{f,s}}{\Delta k_f}, \tag{3.37c}$$

and substituting them into (3.36) yields,

$$\frac{\partial \theta}{\partial \tau} = \frac{\partial^2 \theta}{\partial \eta^2} + \left(\frac{\Delta k_f}{k_f} \right) \frac{\partial \theta}{\partial \eta} \frac{\partial \kappa}{\partial \eta} + \Gamma_1 \frac{\partial^2 \phi}{\partial \eta^2} + \Gamma_1 \left(\frac{\Delta D_{v,eff}}{D_{v,eff}} \right) \frac{\partial \pi}{\partial \eta} \frac{\partial \phi}{\partial \eta}, \tag{3.38}$$

where Γ_1 is defined as the reciprocal of the product of the Stefan and Lewis numbers with quantities specific to Stage I melting,

$$\Gamma_1 = \left(\frac{\lambda_{ig}}{c_f \Delta T} \right) \left(\frac{D_{v,eff} c_f \Delta \rho_v}{k_f} \right) = \left(\frac{1}{St_1} \right) \left(\frac{1}{Le_1} \right). \tag{3.39}$$

Similarly the conservation of mass can be expressed,

$$\frac{\partial \rho_f}{\partial t} = \frac{\partial}{\partial y} \left(D_{v,eff} \frac{\partial \rho_v}{\partial y} \right). \tag{3.40}$$

Applying the chain rule to expand the terms on the r.h.s.,

$$\frac{\partial \rho_f}{\partial t} = D_{v,eff} \frac{\partial^2 \rho_v}{\partial y^2} + \frac{\partial D_{v,eff}}{\partial y} \frac{\partial \rho_v}{\partial y}. \tag{3.41}$$

Recall that the porosity of the frost layer is given by,

$$\varepsilon = \frac{\rho_i - \rho_f}{\rho_i - \rho_a} = \frac{\rho_i - \rho_f}{\Delta \rho_{i,a}}. \tag{3.42}$$

Substituting the dimensionless parameters into (3.41) yields,

$$\frac{\partial \varepsilon}{\partial \tau} = \left(\frac{1}{Le_1} \right) \left(\frac{\rho_f}{\Delta \rho_{i,a}} \right) \frac{\partial^2 \phi}{\partial \eta^2} + \left(\frac{1}{Le_1} \right) \left(\frac{\rho_f}{\Delta \rho_{i,a}} \right) \left(\frac{\Delta D_{v,eff}}{D_{v,eff}} \right) \frac{\partial \pi}{\partial \eta} \frac{\partial \phi}{\partial \eta}. \tag{3.43}$$

The dimensionless boundary conditions are,

$$\left.\frac{\partial \theta}{\partial \eta}\right|_{\eta=0} = -\frac{\delta_f q_s''}{k_f \Delta T} = -Bi_{1,s}, \quad \left.\frac{\partial \phi}{\partial \eta}\right|_{\eta=0} = 0, \quad (3.44a)$$

$$\left.\frac{\partial \theta}{\partial \eta}\right|_{\eta=1} = \left(\frac{\delta_f h}{k_f}\right)\left(\frac{T_a - T_{fs}}{\Delta T}\right) + \left(\frac{\lambda_{ig}}{c_{p,f}\Delta T}\right)\left(\frac{D_{v,eff} c_f \Delta \rho_v}{k_f}\right)\left(\frac{\delta_f h_m}{D_{v,eff}}\right)\left(\frac{\Delta \rho_{v,a,fs}}{\Delta \rho_v}\right)$$

$$= Bi_{1,fs}\left(\frac{\Delta T_{a,fs}}{\Delta T}\right) + \Gamma_1\left(\frac{\delta_f h_m}{D_{v,eff}}\right)\left(\frac{\Delta \rho_{v,a,fs}}{\Delta \rho_v}\right),$$

$$(3.44b)$$

$$\left.\frac{\partial \phi}{\partial \eta}\right|_{\eta=1} = \left(\frac{\delta_f h_m}{D_{v,eff}}\right)\left(\frac{\Delta \rho_{v,a,fs}}{\Delta \rho_v}\right), \quad (3.44c)$$

where the Biot number for Stage I defrost, $Bi_{1,s} = \delta_f h/k_{f,s}$.

Our experiments suggest that porosity does not vary appreciably within the frost layer. Assuming a constant porosity, it can be shown there is no change in the local value of the diffusion coefficient and conductivity from the wall to the frost-air interface. Thus,

$$\Delta D_{v,eff} = D_{v,eff,fs} - D_{v,eff,s} \xrightarrow[\frac{\partial \varepsilon}{\partial y}=0]{} 0, \quad (3.45a)$$

$$\Delta k_f = k_{f,fs} - k_{f,s} \xrightarrow[\frac{\partial \varepsilon}{\partial y}=0]{} 0. \quad (3.45b)$$

Applying (3.45a) and (3.45b) to (3.38) and (3.43), the non-linear terms are eliminated, to give,

$$\frac{\partial \theta}{\partial \tau} = \frac{\partial^2 \theta}{\partial \eta^2} + \Gamma_1 \frac{\partial^2 \phi}{\partial \eta^2}, \quad (3.46)$$

$$\frac{\partial \varepsilon}{\partial \tau} = \left(\frac{1}{Le_1}\right)\left(\frac{\rho_f}{\Delta \rho_{i,a}}\right)\frac{\partial^2 \phi}{\partial \eta^2}. \quad (3.47)$$

It is possible to decouple the relation of temperature to vapor density in (3.47). The assumption is that the temperature of the water vapor is equal to the local frost temperature (local thermal equilibrium). The water vapor density thus varies primarily with the saturation temperature. Furthermore, over small ranges of temperature, vapor density can be approximated by,

$$\rho_v = b_1 T + b_2, \quad (3.48)$$

where the constants are determined from a regression fit of the measurements. Differentiating both sides of (3.48) with respect to y,

$$\frac{\partial^2 \phi}{\partial \eta^2} = b_1 \frac{\Delta T}{\Delta \rho_v} \frac{\partial^2 \theta}{\partial \eta^2}. \tag{3.49}$$

The ratio $\Delta \rho_v / \Delta T$ is the local slope of the vapor pressure-versus-temperature relation. Thus, (3.49) is reduced to,

$$\frac{\partial^2 \phi}{\partial \eta^2} = \frac{\partial^2 \theta}{\partial \eta^2}. \tag{3.50}$$

Applying (3.50) to (3.46),

$$\frac{\partial \theta}{\partial \tau} = (1 + \Gamma_1) \frac{\partial^2 \theta}{\partial \eta^2}. \tag{3.51}$$

Similarly, the change in porosity can be related to changes in temperature by,

$$\frac{\partial \varepsilon}{\partial \tau} = \left(\frac{1}{Le_1}\right) \left(\frac{\rho_f}{\Delta \rho_{i,a}}\right) \frac{\partial^2 \theta}{\partial \eta^2}. \tag{3.52}$$

With the governing equations in dimensionless form, the relative weight of each term in the differential form of the equation can be estimated. In (3.51), it is clear that the diffusion of vapor is an enhancement to the overall heat transfer from the surface, and it is scaled to the Stefan and Lewis numbers.

Recall from (3.2), (3.3a), (3.3b) and (3.9) that the specific heat, thermal conductivity, and diffusion coefficient are proportional to the porosity of the frost layer. Specific heat and thermal conductivity are primarily driven by the properties of ice. Figure 3.6 shows the variability of specific heat, conductivity and diffusion coefficient as a function of porosity at a frost temperature of $-10\ °C$. Applying the physical properties, the Lewis and Stefan numbers as a function of porosity for Stage I defrost are shown in Fig. 3.7. For porosity in the range of 0.4–0.6, $Le_1 \sim 10,000$. This is obvious when evaluating (3.39), as transport due to thermal diffusion (the numerator) is significantly greater than that of mass diffusion (the denominator). The Stefan number is nearly constant for all values of porosity, $St_1 \sim 0.01$, and it is dominated by the large value of heat of sublimation of ice (2604 kJ/kg). The quantity Γ_1, defined as the reciprocal of the product of Stefan and Lewis numbers, is found to vary from zero to greater than unity over the range of porosity. The effect of vapor diffusion can be negligible for extremely small values of Γ_1. From Fig. 3.7, heat transfer due to vapor diffusion is $\sim 1\ \%$ for $\varepsilon = 0.5$, and $<5\ \%$ for $\varepsilon = 0.7$. For such small values, mass transport due to vapor diffusion in Stage I defrost can be assumed negligible for $\varepsilon \sim 0.5$. Applying energy and mass conservation, (3.51) and (3.52) become,

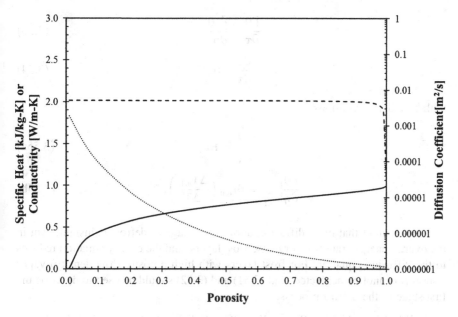

Fig. 3.6 Frost properties as a function of porosity. $T_f = -10°C$. *Dotted line* thermal conductivity, *dashed line* specific heat, *solid line* mass diffusion coefficient

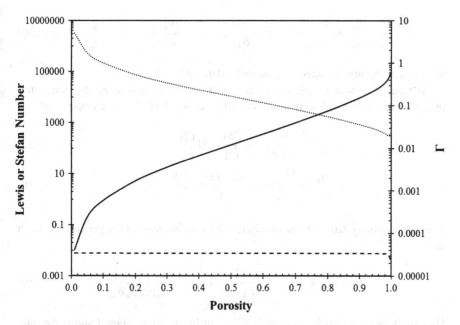

Fig. 3.7 Dimensionless groups for Stage I ($T_f = -10°C$) *Dotted line* lewis number, *dashed line* stefan number, *solid line* gamma

$$\frac{\partial \theta}{\partial \tau} = \frac{\partial^2 \theta}{\partial \eta^2}, \tag{3.53}$$

$$\frac{\partial \varepsilon}{\partial \tau} = 0, \tag{3.54}$$

with boundary conditions,

$$\left.\frac{\partial \theta}{\partial \eta}\right|_{\eta=0} = Bi_{1,s}, \tag{3.55a}$$

$$\left.\frac{\partial \theta}{\partial \eta}\right|_{\eta=1} = Bi_{1,fs}\left(\frac{\Delta T_{a,fs}}{\Delta T}\right). \tag{3.55b}$$

It is apparent that mass diffusion during first stage of defrost is insignificant in the overall energy transport for dense frost layers, and the energy equation reduces to the diffusion equation. For frost layers with high porosity (low density) vapor transport cannot be neglected, and (3.51) and (3.52) should be used to describe the first stage of the defrost process.

Stage II Defrost—Melting-Permeation: Stage II defrost is dominated by the physics of melting, frost front movement toward the surface, and a permeation front moving away from the surface. The relevant dimensionless parameters are,

$$\theta = \frac{T - T_m}{T_a - T_m} = \frac{T - T_a}{\Delta T_{a,m}}, \quad \tau = \frac{u_0 t}{\delta_{f,0}}, \quad M = \frac{m''_{w,p}}{u_0 \varepsilon \rho_w}, \quad \eta = \frac{y}{\delta_{f,0}}, \tag{3.56}$$

where u_0 is the initial convective velocity defined by (3.16).

Applying the dimensionless parameters to the permeation layer, the concentration gradient (3.18) and permeation layer thickness (3.19) become respectively,

$$\frac{\partial S}{\partial \tau} = -\frac{\partial M}{\partial \eta} - U\frac{\partial S}{\partial \eta}, \tag{3.57}$$

$$\frac{\partial \eta_p}{\partial \tau} = \frac{M_{\eta_p=1}}{S_c} + \frac{St_{2,p}Le_{2,p}}{S_c}\left.\frac{\partial \theta}{\partial \eta}\right|_{\eta=1} + U, \tag{3.58}$$

where the velocity ratio, Stefan number, and Lewis number of the permeation layer are,

$$U = \frac{u}{u_0}, \quad St_{2,p} = \frac{c_f \Delta T}{\lambda_{if}}, \quad Le_{2,p} = \frac{k_p}{u_0 \varepsilon c_w \rho_w \delta_{f,0}}. \tag{3.59}$$

Here the Lewis number has a slightly different form than in Stage I and is the ratio of the heat flow due to thermal transport to the bulk movement of the water within the frost layer.

Applying the scaling parameters and the vapor density approximation, the energy equation in the vapor diffusion layer is,

$$\frac{\partial \theta}{\partial \tau} = Le_{2,f}(1 + \Gamma_2)\frac{\partial^2 \theta}{\partial \eta^2} + U\frac{\partial \theta}{\partial \eta}, \tag{3.60}$$

where $\Gamma_2 = (St_{2,v}Le_{2,v})^{-1}$, and

$$St_{2,v} = \frac{c_f \Delta T}{\lambda_{ig}}, \quad Le_{2,v} = \frac{k_f}{D_{v,eff}c_f \Delta \rho_v}, \quad Le_{2,f} = \frac{k_f}{u_0 \varepsilon c_f \rho_f \delta_{f,0}}. \tag{3.61}$$

The boundary conditions are,

$$\left.\frac{\partial \theta}{\partial \eta}\right|_{\eta=0} = -\frac{q_s'' \delta_{f,0}}{k_w \Delta T_{a,m}} = -Bi_{2,s}, \quad S_{\eta=0} = 1, \tag{3.62a}$$

$$\theta_{\eta=\eta_p} = 0, \quad S_{\eta=\eta_p} = S_c, \tag{3.62b}$$

$$\left.\frac{\partial \theta}{\partial \eta}\right|_{\eta=1} = Bi_{2,fs}\left(\frac{T_a - T_{fs}}{\Delta T_{a,m}}\right) + \left(\frac{1}{St_{2,v}}\right)\left(\frac{1}{Le_{2,v}}\right)\left(\frac{\delta_f h_m}{D_{v,eff}}\right)\left(\frac{\Delta \rho_{v,a,fs}}{\Delta \rho_v}\right). \tag{3.62c}$$

For most values of porosity, heat transfer due to vapor diffusion can be neglected (Fig. 3.6). Thus (3.60) can be approximated by,

$$\frac{\partial \theta}{\partial \tau} = Le_{2,f}\frac{\partial^2 \theta}{\partial \eta^2} + U\frac{\partial \theta}{\partial \eta}, \tag{3.63}$$

with boundary conditions,

$$\theta_{\eta=\eta_p} = 0, \tag{3.64a}$$

$$\left.\frac{\partial \theta}{\partial \eta}\right|_{\eta=1} = Bi_{2,fs}\left(\frac{T_a - T_{fs}}{\Delta T_{a-m}}\right). \tag{3.64b}$$

For the case of constant heat flux at the surface, recall from (3.16) that the velocity of the front is related to the heat input at the surface, latent heat of fusion and frost density. If the input heat, q_s'', is held constant, by definition the front velocity will be constant, with $U = 1$. For thin frost layers, surface tension effects prevent draining of the melt liquid from the surface. Assuming the melt water flux is negligible, the permeation mass flux can be approximated as,

$$m_{w,p}'' = \frac{q_s''}{\lambda_{if}\rho_f}\left(\rho_p + \varepsilon \rho_w S_{y=0}\right). \tag{3.65}$$

Stage III Defrost—Dry Out: The non-dimensional parameters for Stage III defrost are,

$$\theta_w = \frac{T_w - T_{w,0}}{T_a - T_{w,0}} = \frac{T_w - T_{w,0}}{\Delta T_{a,w,0}}, \quad \tau = \frac{\alpha_w t}{\delta_{w,0}^2}, \quad \eta_w = \frac{y}{\delta_{w,0}}, \quad (3.66)$$

Applying these groups to the mass and energy (3.30) and (3.34),

$$\frac{\partial \eta_w}{\partial \tau} = \frac{\delta_{w,0}}{\alpha_w} \bar{u}_w + \frac{h_m \delta_{w,0}}{\alpha_w \rho_w} \Delta \rho_{v,a,w}, \quad (3.67)$$

$$\frac{d\theta}{d\tau} = \frac{\rho_w \Delta h_w \delta_{w,0}}{k_w \Delta T_{a,w,0}} \bar{u}_w + \frac{h \delta_{w,0}}{k_w} \theta + \frac{\lambda h_m \delta_{w,0}}{k_w \Delta T_{a,w,0}} \Delta \rho_{v,a-w} + \frac{q_s'' \delta_{w,0}}{k_w \Delta T_{a,w,0}}. \quad (3.68)$$

Thus the rate change of the thickness and temperature of the water film are a coupled set of equations. Both η_w and q_s'' are a function of the film velocity, temperature potential, water vapor density potential, and supplied heat flux. With (3.48) it is possible to expresses vapor density by the water film temperature. The modified equations are,

$$\frac{\partial \eta_w}{\partial \tau} = \frac{\delta_{w,0}}{\alpha_w} \bar{u}_w + \frac{h_m \delta_{w,0}}{\alpha_w \rho_w} b_1 \theta, \quad (3.69)$$

$$\frac{\partial \theta}{\partial \tau} = \frac{\rho_w \Delta h_w \delta_{w,0}}{k_w \Delta T_{a,w,0}} \bar{u}_w + \left[\frac{h \delta_{w,0}}{k_w} + \frac{\lambda h_m \delta_{w,0}}{k_w \Delta T_{a,w,0}} b_1 \right] \theta + \frac{q_s'' \delta_{w,0}}{k_w \Delta T_{a,w,0}}, \quad (3.70)$$

where b_1 is the local slope of the vapor density relation.

From (3.29), the film velocity varies as the square of the film thickness. The film velocity quickly decays to a small value for thin films (Fig. 3.8). Surface tension effects which were neglected in (3.28) retard the film flow. For most practical cases, when the film thickness is small, the film velocity can be assumed to be negligible. Neglecting film velocity and grouping the constants into dimensionless parameters,

$$\frac{\partial \eta_w}{\partial \tau} = -\frac{1}{Le_3} \theta, \quad (3.71)$$

$$\frac{\partial \theta}{\partial \tau} = \left[Bi_{3,w}, + \frac{1}{St_3} \frac{1}{Le_3} \right] \theta + Bi_{3,s}, \quad (3.72)$$

where the dimensionless groups are,

$$Le_3 = \frac{k_w \rho_w}{h_m \delta_{w,0} c_w b_1}, \quad St_3 = \frac{c_w \Delta T_{a,w,0}}{\lambda_{fg}}, \quad Bi_{3,w} = \frac{h \delta_{w,0}}{k_w},$$

$$Bi_{3,s} = -\frac{q_s'' \delta_{w,0}}{k_w \Delta T_{a,w,0}} \quad (3.73)$$

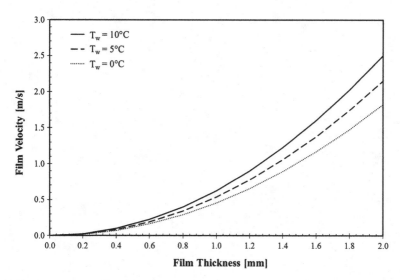

Fig. 3.8 Water film velocity, u

From (3.71), the change in film thickness is driven by the evaporation of the liquid from the surface, and is proportional to the temperature potential. The change in temperature (3.72) is a function of sensible heat transfer from the surface and combined sensible and latent heat transfer at the air-water interface.

References

Aoki K, Hattori M, Ujiie T (1988) Snow melting by heating from the bottom surface. JSME Int J 31(2):269–275

Auracher H (1987) Effective thermal conductivity of frost. In: Proc Int Symp Heat Mass Trans in Refrigeration and Cryogenics, Dubrovnik, 1–5 September 1986, pp 285–302

El Sherbini AI, Jacobi AM (2006) A model for condensate retention on plain-fin heat exchangers. J Heat Transfer 128:427–433

Hao YL, Iragorry J, Tao Y (2005) Frost-air interface characterization under natural convection. J Heat Transfer 127(10):1174–1180

Lee YB, Ro ST (2005) Analysis of the frost growth on a flat plate by simple models of saturation and super saturation. Exp Therm Fluid Sci 29(6):685–696

Na B (2003) Analysis of frost formation in an evaporator. Doctoral dissertation, Pennsylvania State University, University Park

Wiederhold PR (1997) Water vapor measurement: methods and instrumentation. Marcel Dekker, New York

Yan WM, Lin TF (1990) Combined heat and mass transfer in natural convection between vertical parallel plates with film evaporation. Int J Heat Mass Transfer 33(3):529–541

Chapter 4
Experimental Method

Abstract This chapter provides a description of an experimental apparatus constructed to permit real time measurement of frost thickness and planar morphology during growth and melting, as well as heat transfer rates. Quantitative data are obtained via digital reduction of normal and in plane images of the frosted test surface. Data reduction is described, and measurement uncertainties are summarized.

Keywords Frost growth • Defrost • Frost thickness • Optical measurement • Digital analysis

Nomenclature

A	Area (m^2)
c	Specific heat (J/kg K)
C	Thermal capacitance (J/K)
d	Droplet diameter (m)
E	Energy (J)
h	Heat transfer coefficient (W/m^2 K)
h_m	Mass transfer coefficient (m/s)
i	Enthalpy (J/kg)
k	Thermal conductivity (W/m K)
L	Length (m)
m	Mass (kg)
$\dot{m}$	Mass transfer (kg/s)
m''	Mass flux (kg/m^2 s)
N	Number (–)
p	Pressure (N/m^2)
P	Power (W), perimeter (m)
q''	Heat flux (W/m^2)
Q	Heat transfer (W)
r_c	Radius of curvature (m)

© Springer International Publishing Switzerland 2015
W.F. Mohs, F.A. Kulacki, *Heat and Mass Transfer in the Melting of Frost*,
SpringerBriefs in Applied Sciences and Technology,
DOI 10.1007/978-3-319-20508-3_4

43

R   Ideal gas constant (J/kg K)
R_{th} Thermal resistance (K/w)
t   Time (s)
T   Temperature (K)
U   Overall heat transfer coefficient (w/m^2 K)
V   Volume (m^3)
$\dot{V}$ Volumetric flow (m^3/s)

Greek Symbols

δ Thickness, height (m)
ε Porosity (–)
η Efficiency
λ_{if} Latent heat of fusion (J/kg)
ρ Density (kg/m^3)

Subscripts

a   Air
c   Calibration, characteristic
ch   Chamber
cp   Cold plate
d   Defrost, droplet
e   Edge
f   Frost
hu   Humidifier
i   Ice
l   Liquid
o   Outlet
s   Surface
sp   Heat flux spreader
t   Total
ts   Test surface
TE   Thermoelectric
v   Vapor
w   Water
wall   Wall
∞ Exterior

4.1 Apparatus

There is a gap in knowledge of the effects of frost morphology on the defrost process. The majority of the experiments on defrosting have been conducted at the system level (Al-Mutawa et al. 1998a, b, c, d; Al-Mutawa and Sherif 1998; Muehlbauer 2006; Donnellan 2007). These investigations have characterized the performance cost of defrost on system operation. An important finding is that defrost efficiency, which is dependent on frost morphology, is related to the growth conditions of the frost layer. Conversely, there apparently have been no investigations of the effect of frost morphology on the defrost process. We address this gap via controlled measurements of heat transfer, transient frost thickness in frosting and defrosting, and transient planar morphology on a thermally controlled vertical surface. High resolution images of the frost permit non-invasive measurement of frost thickness and planar morphology, and such images are digitally reduced to quantitatively characterize the frost and defrost processes. These data provide a basis for analysis and development of semi-empirical models and comparison to the multi-stage defrost model developed in Chap. 3.

Figure 4.1 shows the apparatus and instrumentation. Its main elements are the test chamber, test surface, humidity generator, hygrometer, chiller, digital video cameras with zoom microscopes, pressure transducer, and data acquisition system. The test chamber is capable of controlling ambient temperatures in the range of

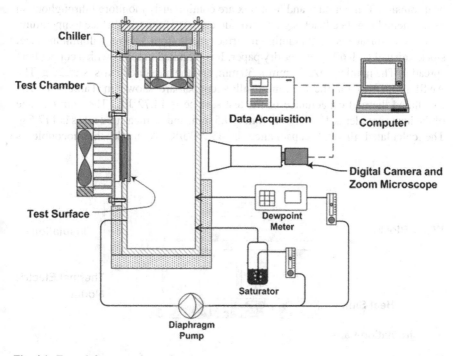

Fig. 4.1 Frost-defrost experimental apparatus

10 to −15 °C at 1 atm total pressure, chamber dew point from 0 to −20 °C, and test surface temperatures to −20 °C. In the paragraphs below, we describe the key design elements of the apparatus and the data reduction methodology. More detail can is presented by Mohs (2013) and Janssen (2011).

The test chamber is a well-insulated 12 mm thick polycarbonate enclosure housing the test surface. The chamber isolates the test surface from the ambient space and prevents air and moisture exchange. The interior dimensions of the chamber are 20 cm high × 10 cm wide × 5 cm deep. The interior volume of the chamber is 1050 cm³, and neglecting the space occupied by the chiller, the net interior volume is 980 cm³. All access points into the chamber are sealed to limit the ingression of air and moisture, and upon evacuation to 500 mbar, the return to 1 atm requires ~10 min.

The chiller (Fig. 4.2) is a heat sink in series with two Peltier thermoelectric modules. Heat is removed by a cold plate using the laboratory water supply as the cooling fluid. Circulation fans fixed to the heat sink provide 20 air exchanges/min at a mean velocity of 0.9 m/s. The temperature of the chiller is measured by several thermocouples mounted within the heat sink and is controlled by varying its applied voltage. During frost growth the temperature of the cold plate is maintained above the dew point to avoid condensation and frost formation. The test surface is mounted on a Peltier thermoelectric module in series with a copper heat spreader and heat sink (Fig. 4.3). Two embedded thermocouples measure the temperature near the frosting surface in the center and near a corner. A heat flux sensor measures heat transfer. Temperature and heat flux are continuously monitored throughout an experimental run. Feedback signals provide control of the test surface temperature.

The test surface is a flat aluminum surface made from 3.88 mm aluminum sheet stock finished with 600 grit wet-dry paper. It is mounted in series with a copper heat spreader. The nominal size 38 mm × 38 mm, and the measured mass is 15.29 g. The ANSI properties for the aluminum alloy copper are shown in Table 4.1. The calculated thermal capacitance of the test surface is 14.72 J/K. The nominal size of the heat spreader is 45 mm × 40 mm × 9.5 mm, and its measured mass is 117.5 g. The calculated thermal capacitance is 45.26 J/K. A single thermocouple is

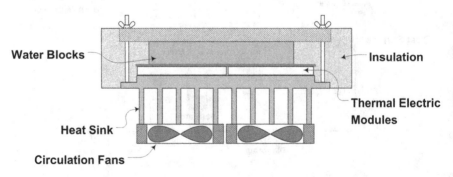

Fig. 4.2 Chiller assembly

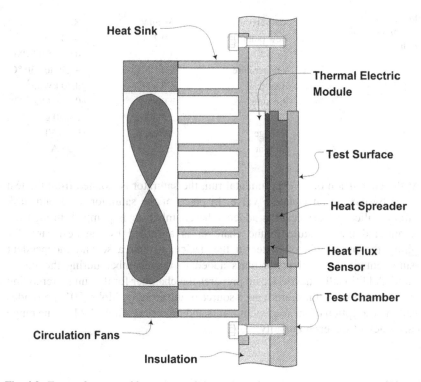

Fig. 4.3 Test surface assembly

Table 4.1 Material properties of test surface assembly

Property	Aluminum 5052-H32	Copper D110
Density (kg/m^3)	2680	8933
Thermal conductivity (W/m K)	138	388
Specific heat (J/kg K)	936	385

embedded at its center. All heat conducting interfaces in Fig. 4.3 are coated with a conductive grease to minimize contact resistances.

The humidity generator (Fig. 4.1) is based on the NIST frost-point generator (Scace et al 1997). Air is drawn from the test chamber into the humidifier by an oil less diaphragm air pump. The flow stream is split into two paths. One stream is directed to the saturator where moisture is absorbed by the air stream. The other air stream bypasses the saturator and is directed to the hygrometer to measure dew point temperature. The flow to saturator and hygrometer are controlled by variable flow rotameters. By controlling the amount of bypass airflow and the temperature of the saturator, a precise level of moisture can be returned to the test chamber. The mass of saturator with the charge of distilled water is measured at the start of an experimental run and again at the end of the run, and the difference is the mass of water added to the test chamber.

Table 4.2 Summary of instrument accuracy and range

Measurement	Accuracy	Range
Dew point	±0.2 °C	−50 to 90 °C
Flow	±5 %	0.5 to 5.0 LPM
Temperature	±0.5 °C	−250 to 350 °C
Heat flux	±0.5 %	±30 kW/m^2
Pressure	±0.05 %	0–210 kPa
Mass	±3 mg	0–320 g
Voltage	±0.03 %	0–10 VDC
Current	±.25 %	±25 A

At the conclusion of an experimental run, the saturator is isolated from the test chamber by closing an isolation valve. In place of the saturator, a second flask containing silica gel desiccant is added. The chamber air is pumped through the desiccant and the moisture in the chamber is absorbed by the desiccant. By weighing the mass gain of water in the desiccant, and a second independent measurement of the total water mass added to the chamber during the run is captured. Additionally, the desiccant prepares the chamber for the run by removing any residual moisture, thus removing a source of uncertainty. Mohs (2013) provides a detailed description of measurement limits and precision. Table 4.2 lists the range and accuracy of the each quantity.

4.2   Characterization of the Test Chamber

Heat transfer, heat transfer coefficients, mass balances etc. are essential to the overall accuracy of results for frost properties and defrost efficiency. These quantities form the basis of validation of the theory described in Chap. 3.

Chamber Heat Balance: The heat balance for the test chamber is,

$$Q_{ts} + Q_{ch} + Q_{cp} + Q_{hu} = 0, \tag{4.1}$$

where Q_{ts} is the heat extracted by the test surface, Q_{ch} is the heat leakage into the test chamber, Q_{cp} is the heat extracted by the chiller, and Q_{hu} is the heat added by the humidifier. Heat extracted by the test surface is measured by the heat flux sensor. Heat leakage from the chamber is determined by multiplying the overall chamber heat transfer coefficient, UA_{ch}, by the temperature difference between the interior and exterior of the chamber. The overall heat transfer coefficient determined by calibration test is,

$$UA_{ch} = \frac{Q_{ch,c}}{(T_\infty - T_{ch})}, \tag{4.2}$$

where $Q_{ch,c}$ is the sensible heat added to the chamber during calibration, and T_∞ and T_{ch} are the average exterior and interior air temperatures respectively. Extracted heat is estimated by the input power, P, and heat pumping efficiency of the thermoelectric module, η_{TE},

$$Q_{cp} = P \cdot \eta_{TE}. \tag{4.3}$$

Heat addition by the humidifier is calculated by the change in air enthalpy across the device,

$$Q_{hu} = \dot{V}_a \rho_a (i_{a,o} - i_{a,i}), \tag{4.4}$$

where $\dot{V}_a$ is the volumetric flow rate, ρ_a is the average density, and $i_{a,o}$ and $i_{a,i}$ are the outlet and inlet moist air enthalpies respectively.

Water Mass Balance: The overall mass balance of water in the test chamber can be expressed,

$$m_{l,hu} = m_{f,ts} + m_{v,ch}, \tag{4.5}$$

where $m_{l,hu}$ is the total mass of water vapor introduced by the humidifier during the test, $m_{f,ts}$ is the mass of the frost adhered to the test surface, and $m_{v,ch}$ is the mass of the water vapor in the air of the chamber. The estimated mass of the frost is,

$$m_{f,ts} = \rho_i V_{f,i} = \rho_i V_{f,t}(1 - \varepsilon), \tag{4.6}$$

where ρ_i is the density of ice, $V_{f,i}$ is the volume of the ice in the frost layer that is related to the total volume of frost, $V_{f,t}$ and by the frost porosity, ε. The total volume of the frost layer is determined by the product of the test surface area, A_{ts} and the frost average thickness, δ,

$$V_{f,t} = A_{ts}\delta. \tag{4.7}$$

The mass of the water vapor in the chamber is determined by the ideal gas law,

$$m_{v,ch} = \frac{p_v V_{ch}}{RT_{ch}} \tag{4.8}$$

where p_v is the vapor pressure of water in the chamber, V_{ch} is the volume of the chamber, T_{ch} is the chamber temperature, and R is the ideal gas constant for water vapor. The mass transfer rate is estimated as the change in the water vapor mass over the duration of an experimental run,

$$\dot{m}_{v,ch} = \frac{\Delta m_{v,ch}}{\Delta t}. \tag{4.9}$$

Test Surface Energy Balance: The defrost process is inherently transient, and it is necessary to determine the net heat flow at the test surface to complete the analysis. The heat flow at the test surface, Q_{ts}, is a function of the heat flow into the heat

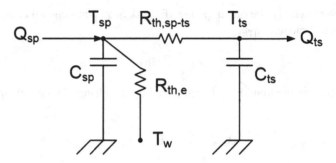

Fig. 4.4 Lumped capacitance model of test surface

spreader, Q_{sp}, the heat flow at the edges, Q_e, and the internal energy changes of the test surface and heat spreader. A lumped analysis approach is developed to simplify the governing equations. For a lumped analysis, temperature variations within the body are neglected, and $Bi < 0.1$. The heat spreader and test surface have a small characteristic length (2.6 and 1.6 mm respectively) and high thermal conductivities (388 and 138 W/m K respectively). As long as the heat transfer coefficient is below 8000 W/m^2 K, a lumped analysis will be valid. Figure 4.4 is the equivalent resistance-capacitance circuit for the heat spreader and test surface.

The transient energy balance at node T_{sp} is,

$$\frac{dT_{sp}}{dt} = \frac{1}{C_{sp}}\left[\frac{1}{R_{th,sp,ts}}(T_{ts} - T_{sp}) + \frac{1}{R_{th,e}}(T_w - T_{sp}) + Q_{sp}\right], \qquad (4.10)$$

where $C_{sp} = 45.26$ J/K and is determined from the materials properties of spreader, and T_w is the temperature of the chamber wall behind the heat flux spreader. The thermal resistance, $R_{th,sp,ts}$, is defined as the contact resistance between the heat spreader and test surface. The edge thermal resistance, $R_{th,e}$, is the resistance between the heat spreader and the chamber walls. The mean wall temperature, T_w, is assumed to be the average of the inner and outer chamber temperatures. The thermal resistances, $R_{th,sp,ts}$ and $R_{th,e}$ are determined through steady-state test. Similarly, the energy balance at node T_{ts} is,

$$\frac{dT_{ts}}{dt} = \frac{1}{C_{ts}}\left[\frac{1}{R_{th,sp,ts}}(T_{sp} - T_{ts}) + Q_{ts}\right], \qquad (4.11)$$

where C_{ts} is the thermal capacitance of the test surface and is calculated to be 14.72 J/K. The heat transfer coefficient at the surface is defined,

$$h_{ts} = \frac{Q_{ts}}{A_{ts}(T_{ch} - T_{ts})}. \qquad (4.12)$$

Defrost Efficiency: Defrost efficiency is defined as the ratio of the minimum energy required to melt the frost layer to the total energy supplied to melt the frost, $\eta_d = E_f/E_d$.

The minimum energy to melt the frost, E_f, is the sum of the latent and sensible heat of the frost layer. The defrost energy, E_d, is the time integrated heat flow during the defrost process. In practice the supplied energy is approximated by the numerical integration of the measured heat flux, q_{ts}'', over the recording interval, Δt. The defrost efficiency during the defrost test is calculated by,

$$\eta_d = \frac{m_f(\lambda_{if} + c_f \Delta T)}{\sum A_{ts} q_{ts}'' \Delta t}. \qquad (4.13)$$

4.3 Visual Measurement and Quantification

Digital cameras with zoom microscopes are used to capture normal and in plane images of frost growth and defrost processes. The cameras are positioned outside the test chamber and access ports allow for viewing of the test surface. The test surface is illuminated with a fiber optic light ring and light source attached to the normal view camera lens and provides the necessary illumination for the side camera as well. Normal images are captured by a 5.1 megapixel CMOS camera with a maximum magnification of ×4.5 and resolution of 1 μm at the maximum magnification and resolution. Table 4.3 summarizes the image resolution. The in plane profile of the frost is captured by a second camera with a zoom imaging lens. The camera has a 1.27 cm (0.5 in.) monochrome sensor capable of a resolution of 1280×1024 at frame rates of 25 s^{-1}. The camera is capable of capturing higher frame rates (up to 50 s^{-1}) with reduced resolution. The zoom imaging lens has a maximum magnification of ×6.0 and minimum field of view of 6.4 mm. The maximum image resolution is 133 px/mm, which allows for resolving frost structures of 15 μm. The resolution of both cameras was verified using a calibrated test target based on the 1951 USAF MIL-STD-150A.

From analysis of the images, it is possible to determine a number of physical characteristics of the frosting and defrosting process. These characteristics are sometimes used for indirect calculations. For example, the porosity of the frost is needed to calculate the mass of the ice at the beginning of the defrost process. Other readily determined from the change in the frost front location. The mass transfer of the water vapor during the dry out stage can be seen as a reduction in the volume of the droplets.

Table 4.3 Image resolution (px/mm and μm) for frontal camera and lens

| Magnification | Camera resolution (pixels) | | | | | | | |
| | 2592×1944 | | 1280×960 | | 1024×768 | | 640×480 | |
	Camera	Lens	Camera	Lens	Camera	Lens	Camera	Lens
×1	406	4.92	201	9.97	201	9.97	100	19.94
×2	765	2.61	378	5.29	378	5.29	189	10.59
×3	1308	1.53	646	3.10	646	3.10	323	6.19
×4	1753	1.14	866	2.31	866	2.31	433	4.62
×4.5	2112	0.95	1043	1.92	1043	1.92	521	3.84

Porosity: Assuming that the ice crystals within the frost layer are uniformly dispersed in the frost layer, porosity can be determined along any two-dimensional plane,

$$\varepsilon = 1 - \frac{V_i}{V_t} = 1 - \left(\frac{A_i}{A_t}\right)\left(\frac{L_i}{\delta}\right) \approx 1 - \frac{A_i}{A_t}, \qquad (4.14)$$

where A_i is the area occupied by the ice crystals and A_t is the total area and L_i is their characteristic length. Assuming that, L_i, ≈ 1, porosity can be approximated by the ratio of the occupied area to the total area.

Porosity can be estimated through a visual technique. As seen in Fig. 4.5, the in plane ice crystals can be enhanced against the in plane voids. By gridding the enhanced photo and counting the occupied squares, N_i, and dividing by the total number of squares, N_t, the porosity is,

$$\varepsilon \approx 1 - \frac{N_i}{N_t}. \qquad (4.15)$$

To automate and improve the accuracy of the calculation, the digital photo can be evaluated at the pixel level. The image is converted to a binary black and white image, and porosity is calculated by counting the number of pixels that are white (pixel value of 1) and dividing by the total number of pixels. The method is calibrated through a process where the accumulated mass of the ice is measured directly.

Frost Thickness: From in place images, it is possible to capture the time rate of change of the frost thickness during frost growth and defrost. Each image is converted into a black and white image, and pixel location of the frost-air front is

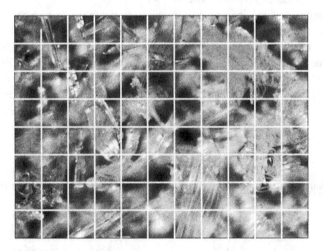

Fig. 4.5 Enhanced frost image

Fig. 4.6 Idealized water droplet on surface

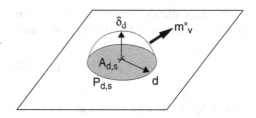

determined by using the ratio of pixels to spatial dimension of the photo (see Janssen 2011 for details).

Droplet Area and Volume: To determine the water vapor mass transfer during the final stage of the defrost process it is necessary to determine the size, shape, and distribution of water droplets remaining on the surface at the end of the melt process (Fig. 4.6). The mass flux of water vapor leaving the droplets surface is,

$$m_v'' = \frac{\partial}{\partial t}\left(\frac{m_d}{A_d}\right), \tag{4.16}$$

where m_d is the mass of a droplet, and A_d is the surface area of the droplet. Assuming that the water density varies minimally over time, it can be moved outside of the derivate,

$$m_v'' = \rho_w \frac{\partial}{\partial t}\left(\frac{V_d}{A_d}\right). \tag{4.17}$$

Assuming the droplets can be modeled as a spherical dome, the surface area and volume are,

$$A_d = 2\pi r_c \delta_d, \tag{4.18}$$

$$V_d = \pi \delta_d \left(r_c \delta_d - \frac{1}{3}\delta_d^2\right), \tag{4.19}$$

where δ_d is the height of the droplet, and r_c is the radius of curvature which is a function of droplet height and equivalent diameter, d,

$$r_c = \frac{\delta_d^2 + (d/2)^2}{2\delta_d}. \tag{4.20}$$

Water droplets generally have an irregular shape. An equivalent droplet diameter is therefore calculated as the hydraulic diameter at the solid-liquid interface,

$$d = \frac{4A_{d,s}}{P_{d,s}}, \tag{4.21}$$

where $A_{d,s}$ is the contact area and $P_{d,s}$ is the perimeter of the droplet at the solid surface interface. The characteristic diameter is defined as the ratio of the droplet volume to surface area,

$$d_c \equiv \frac{V_d}{A_d} = \frac{r_c \delta_d - \frac{1}{3}\delta_d^2}{2r_c}. \tag{4.22}$$

The characteristic diameter is a function of the droplet height, contact area, and perimeter. Using time-lapse photography, these three factors can be determined. The droplet height is measured directly from the side profile photographs. The droplet contact area and perimeter are determined from the normal photographs. The images are enhanced to provide a clear boundary of the droplet. The mass flux is determined from the rate of change of the characteristic diameter,

$$m_v'' = \rho_w \frac{\partial}{\partial t}(d_c) \approx \rho_w \frac{\Delta d_c}{\Delta t}. \tag{4.23}$$

Finally with vapor mass flux, the mass transfer coefficient, $h_{m,v}$, can be determined from,

$$h_{m,v} = \frac{m_v''}{(\rho_{v,a} - \rho_{v,s})}, \tag{4.24}$$

where $\rho_{v,a}$ and $\rho_{v,s}$ are the water vapor density of the free stream air and saturated liquid interface respectively.

Prior to obtaining frost growth and defrost data, care was taken to fully characterize the experimental apparatus, especially the heat transfer surface assembly. Calculations and measurements for dry operation were run, and heat transfer coefficients on the surface and the transient response of the surface assembly (surface-heat spreader-heat sink) were determined. Very good agreement of measurement with a lumped capacitance-resistance model was obtained. Details of the measurements and modeling are given by Mohs (2013).

4.4 Experimental Uncertainty

Table 4.4 summarizes the total experimental uncertainty of quantities and dependent variables used in the reduction of the data. Details are given by Mohs (2013).

Table 4.4 Total experimental uncertainties

Quantity	Total uncertainty (%)
Chamber water vapor mass, $m_{v,ch}$	±2.5
Frost mass, $m_{f,ts}$	±5
Thermal resistances, R_{th}	±2
Test surface heat flow, Q_{ts}	±8
Heat transfer coefficient, h_{ts}	±8
Defrost efficiency, η_d	±12
Frost height, δ_f	±5
Frost porosity, ε	±10
Melt velocity, u	±8
Droplet height, δ_d	±5
Vapor mass flux, m_v''	±6
Vapor mass transfer coefficient, $h_{m,v}$	±8

References

Al-Mutawa NK, Sherif SA (1998) Determination of coil defrosting loads: part V—analysis of loads (RP-622). ASHRAE Trans 104:344–355

Al-Mutawa NK, Sherif SA, Mathur G (1998a) Determination of coil defrosting loads: part III—testing procedures and data reduction (RP-622). ASHRAE Trans 104:303–312

Al-Mutawa NK, Sherif SA, Mathur GD, West J, Tiedeman JS, Urlaub J (1998b) Determination of coil defrosting loads: part I—experimental facility description (RP-622). ASHRAE Trans 104:268–288

Al-Mutawa NK, Sherif SA, Steadham JM (1998c) Determination of coil defrosting loads: part IV—refrigeration/defrost cycle dynamics (RP-622). ASHRAE Trans 104:313–343

Al-Mutawa NK, Sherif SA, Mathur GD, Steadham JM, West J, Harker RA, Tiedeman JS (1998d) Determination of coil defrosting loads: part II—instrumentation and data acquisition systems (RP-622). ASHRAE Trans 104:289–302

Donnellan W (2007) Investigation and optimization of demand defrost strategies for transport refrigeration systems. Doctoral dissertation, Galway-Mayo Institute of Technology, Galway

Janssen DD (2011) Experimental strategies for frost analysis. Master's thesis, University of Minnesota, Minneapolis

Mohs WF (2013) Heat and mass transfer during the melting process of a porous frost layer on a vertical surface. Doctoral dissertation, Universality of Minnesota, Minneapolis

Muehlbauer J (2006) Investigation of performance degradation of evaporators for low temperature refrigeration applications. Master's thesis, University of Maryland, College Park

Scace GE, Huang PH, Hodges JT, Olson DA, Whetstone JR (1997) The new NIST low frost-point humidity generator. In: Proceedings of the 1997 NCSL workshop and symposium, Atlanta

Chapter 5
Measurement of the Defrost Process

Abstract Quantitative and visual data on defrost are presented. The data base comprises normal and in plane images of the defrost process over a range of ambient temperature, dew point, and surface temperature. Twelve frost layers are created at prescribed surface temperature, dew point, and ambient temperature. Melting is initiated by application of heating on the frosted surface. Predictions of the multistage defrost model developed in Chap. 3 are compared to the reduced data where possible, and empirically based relations for heat and mass transfer are developed. Assumptions used to simplify the differential equations for coupled heat and mass transfer in Chap. 3 are validated by the measurements. Overall defrost efficiency is proportional to initial frost thickness.

Keywords Frost profile • Frost growth • Defrost • Frost properties • Heat transfer • Mass transfer • Defrost efficiency

Nomenclature

A	Area (m^2)
c	Specific heat (J/kg K)
E	Energy stored (J)
f	Wetted area fraction, A_w/A_s
h	Heat transfer coefficient (W/m^2 K)
h^*	Total heat transfer coefficient (5.11)
h_m	Mass transfer coefficient (m/s)
Le_1	Lewis number for Stage I (3.39)
Le_{2f}	Lewis number for frost in Stage II (3.61)
Le_{2v}	Lewis number for vapor in State II (3.61)
m	Mass (kg)
m''	Mass flux (kg/m^2 s)
Q	Heat transfer (W)
q''	Heat flux (W/m^2)

© Springer International Publishing Switzerland 2015

W.F. Mohs, F.A. Kulacki, *Heat and Mass Transfer in the Melting of Frost*, SpringerBriefs in Applied Sciences and Technology, DOI 10.1007/978-3-319-20508-3_5

St_1 Stephan number for Stage I (3.39)
St_{2v} Stephan number for vapor in State II (3.61)
t   Time (s)
T   Temperature (K)

Greek Symbols

δ Frost thickness (m)
ε Porosity (–)
λ_{fg} Latent heat of vaporization (J/kg)
λ_{if} Latent heat of fusion (J/kg)
Γ_1 $(St_1Le_1)^{-1}$
Γ_2 $(St_{2v}Le_{2v})^{-1}$
ρ Density (kg m^3)

Subscripts

0   Initial time
1,2,3   Denotes defrost stage
ch   Chamber
d   Defrost
dp   Dew point
f   Frost
fg   Evaporation
i   Ice
lt   Latent
m   Melt
s   Surface
sn   Sensible
ts   Test surface
v   Vapor
w   Water, wetted

5.1   Scope of the Measurements

The frost-defrost cycle comprises a growth phase at given ambient temperature, dew point and surface temperature. When frost thickness has reached its maximum steady state, melting is initiated. The defrost process continues through the dry out stage to the point at which isolated liquid droplets are seen. The data set comprises normal and in plane images during the melting processes. The visual data are digitally reduced and are used to develop general relations for several frost properties.

Frost porosity, density and thermal conductivity are affected by the shape and compactness of the ice crystals, and prior research shows that controlling factors are the bulk air temperature and degree of super-saturation. Super saturation is measured by the temperature difference between the ambient dew point and surface temperature. In the low temperature range, 0 to −20 °C, the primary ice crystalline structures are needles, dendrites, and plates. Factors that affect the frost growth rate are the difference between the ambient and surface temperatures and super saturation, ambient humidity, and surface temperature. The dew point is always bounded by the ambient and surface temperatures ($T_{ch} > T_{dp} > T_{ts}$). The ambient temperature range of interest is 0 to −20 °C, and super saturation ranges from 0 to 15 °C. Table 5.1 summarizes nominal temperatures of the experiments, and Table 5.2 shows the test conditions for the data base.

The general characteristics of frost growth and defrost are shown in Figs. 5.1 and 5.2 respectively. Figure 5.1 shows the test surface heat flux, average chamber ambient temperature, dew point, and surface temperature during a typical frost formation process (see Mohs 2013 for the full data base). Humidity is added to the chamber ambient at ~10 min into each experiment. The frost growth period is ~4 h,

Table 5.1 Conditions in the frost growth experiments

Ambient temperature (°C)	Ambient dew point (°C)	Surface temperature (°C)	Super saturation (°C)	Surface/air temperature difference (°C)
0	−5	−10	5	−10
0	−10	−20	10	−20
−5	−10	−15	5	−10
−5	−15	−20	5	−15
−10	−15	−20	5	−10
−10	−15	−25	10	−15

Table 5.2 Summary of frost conditions at the end of frost growth

Experiment	$T_{ch}/T_{dp}/T_{ts}$ (°C)	Frost thickness (mm)	Porosity (−)	Mass (g)
1	−8.5/−19.0/−20.2	1.08	0.54	0.68
2	−1.2/−8.5/−10.0	0.76	0.58	0.43
3	0.0/−8.3/−19.2	2.68	0.42	2.13
4	−7.9/−18.1/−19.4	1.82	0.48	1.31
5	−0.6/−8.4/−9.6	0.73	0.54	0.45
6	−7.2/−16.8/−19.5	1.75	0.51	1.16
7	−5.6/−13.2/−18.9	2.03	0.49	1.41
8	−5/−12.5/−18.4	1.67	0.53	1.08
9	−8.6/−16.2/−18.9	1.17	0.47	0.84
10	−4.7/−13.9/−17.8	0.69	0.54	0.43
11	−4.6/−12.1/−18.8	1.59	0.43	1.25
12	−3.4/−10.0/−17.5	1.63	0.45	1.19

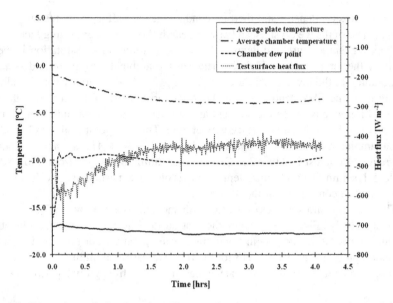

Fig. 5.1 Temperatures during frost formation test. Average temperatures are: surface, −3.4 °C; dew point, −10 °C, and ambient, −17.6 °C (Experiment No. 12, Table 5.2)

while a complete defrost takes on the order of minutes. During defrosting, the applied voltage to the thermoelectric cell is held constant. When Stage II defrost is reached, a constant surface temperature is seen, i.e., the small plateau in the temperature trace, with a corresponding heat flux peak as the phase transition to liquid takes place. Once the phase transition is complete, heating continues to increase the surface temperature due to sensible heat transfer across the water film on the test surface and the frost layer. The next transition occurs when frost completely melts, leaving only liquid film and droplets on the surface. The temperature continues to rise as heat is applied to the surface, and the water droplets evaporate. The highest heat flux is seen during the melting stage. The heat flux during the dry out stage is nearly constant.

5.2   Frost Growth

Figure 5.3 shows in plane and normal images of the frost layer during a typical growth phase. The first image is taken ~10 min after humidity is added to the chamber, and the interval between the images is ~20 min. Frost crystals quickly grow from nucleation sites on the surface (Na 2003) and form as flat plates (Fig. 4.5). Accelerated growth at crystal tips is driven by the high concentration gradient of water vapor at the frost-air interface. As the crystal grows and moves away from the surface, the internal temperature gradient causes the local vapor

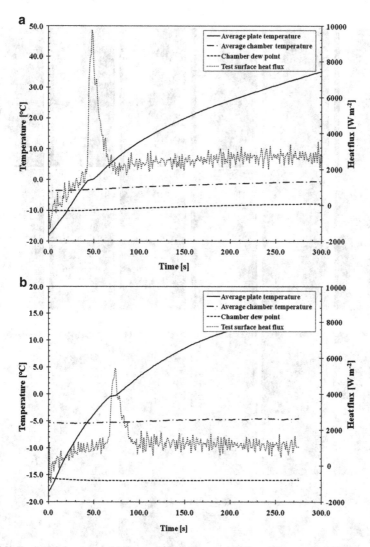

Fig. 5.2 Typical defrost data. (**a**) $T_{ts} = -3.4\ °C$, $T_{dp} = -10\ °C$, and $T_{ch} = -17.5\ °C$ (Experiment No. 12, Table 5.2). (**b**) $T_{ts} = -8.6\ °C$; $T_{dp} = -16.2\ °C$, and $T_{ch} = -18.9\ °C$ (Experiment No. 9, Table 5.2)

pressure at the tip to be closer to that of the ambient air, which causes a reduction of water vapor concentration and reduces the rate of tip growth. The ice plates increase in thickness and length until neighboring plates touch to form a new nucleation site. Most growth occurs during the first 60 min when ice crystals advance rapidly away from the surface. At the later stages of growth, the frost layer comprises larger ice crystals and a more dense structure, which indicates a reduced internal porosity. Frost growth slows as the tip temperature approaches the ambient dew point.

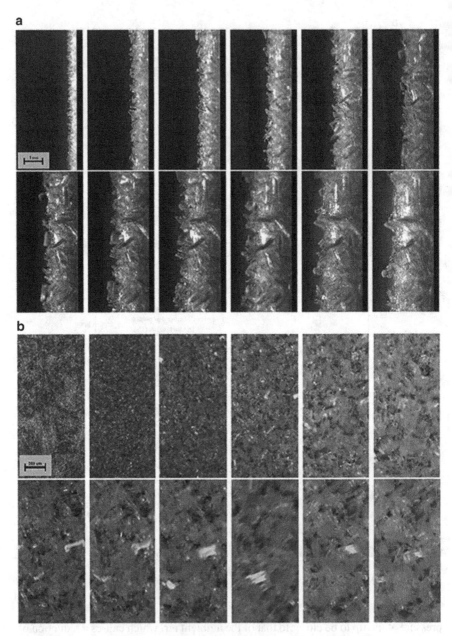

Fig. 5.3 Frost thickness profiles (**a**) and corresponding normal images (**b**) at 20 min intervals. $T_{ch} = 0\ °C$, $T_{dp} = -8.3\ °C$, and $T_{ts} = -19.2\ °C$ (Experiment No. 3, Table 5.2)

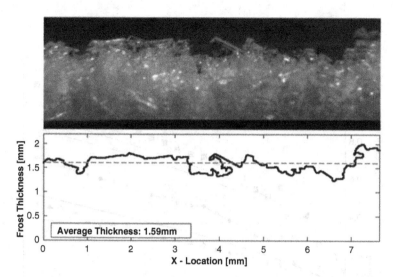

Fig. 5.4 Comparison of digital image and edge capture (Janssen 2011; Janssen et al. 2012a, b)

Once this point is reached, crystal growth is dominated by an increase in the overall size of ice crystals. The average porosity of the frost layer decreases as time progresses.

In plane images of frost thickness are digitally converted to produce an edge profile (Fig. 5.4). The edge profile is arithmetically averaged to yield the reported frost thickness. Figure 5.5 shows frost growth for $T_{ch} = -1.2$ °C, $T_{dp} = -8.5$ °C, and $T_{ts} = -10.0$ °C. Janssen et al. (2012a, b) find that some models of frost growth correlate well to maximum measured thickness, while others agree more closely with the minimum measured thickness. The inconsistency maybe due to differences in either the technique used to measure frost layer thickness or the estimation of the frost surface temperature which is a derived value. The model proposed by Lee and Ro (2002) gives the best agreement to the data.

The porosity of the frost layer is determined from the normal images of the frost layer. The contrast of the image is enhanced to aid in detection of ice crystals and voids and he image is converted to a binary black and white image (Fig. 5.6). Pixels occupied by ice (white, +1 value) are counted, and the porosity is calculated by dividing the occupied pixels by the total pixel count as described in Chap. 4. Computed frost porosities at the conclusion of the frost growth phase of each experiment are shown in Table 5.2. Figure 5.7 shows the change in frost porosity, thickness, and mass during growth for $T_{ch} = 0.0$ °C, $T_{dp} = -8.3$ °C, and $T_{ts} = -19.2$ °C. Porosity is initially unity but decreases as the layer grows. From thickness and porosity, the accumulated mass of the frost layer is given by,

$$m_f = A_{ts}\delta_f(1 - \varepsilon)\rho_i, \tag{5.1}$$

where A_{ts} is the area of the test surface, and ρ_i is the density of ice (920 kg/m^3). Porosity is linearly proportional to thickness (Fig. 5.8), and for present data base and frost growth conditions,

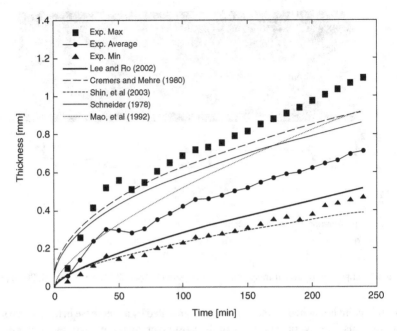

Fig. 5.5 Frost growth rate (Janssen et al. 2012a, b)

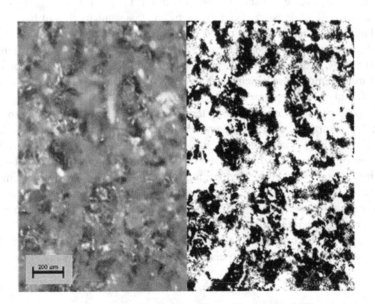

Fig. 5.6 Normal image and enhanced image (Experiment No. 3, Table 5.2)

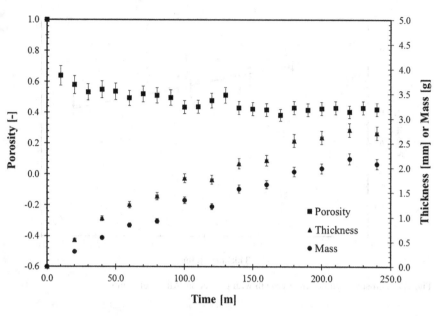

Fig. 5.7 Frost porosity and thickness during growth (Experiment No. 3, Table 5.1)

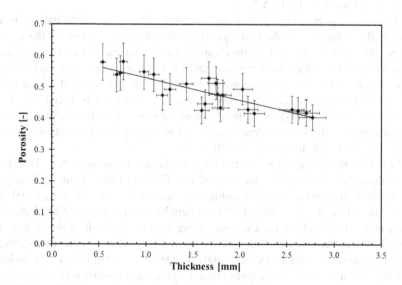

Fig. 5.8 Frost porosity as a function of thickness (Experiment No. 3, Table 5.2)

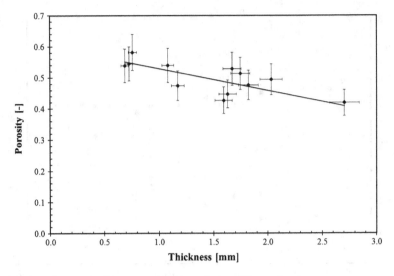

Fig. 5.9 Porosity at end of frost growth with prediction. All experiments

$$\varepsilon = -0.0701\delta_f + 0.5992. \tag{5.2}$$

Porosity at the conclusion of growth is shown in Fig. 5.9 for the 12 growth experiments listed in Table. 5.2.

Table 5.2 also lists the accumulated frost mass based on weight measurement described in Chap. 4. The thickest layer (Experiment No. 3) is grown at the largest temperature differences between the chamber ambient air and surface temperatures. This layer also has the lowest porosity and the most accumulated mass (2.13 g). Frost layers with the highest porosity are grown at the smallest temperature difference between the chamber dew point and test surface temperatures. This result implies that a good estimate of porosity can be made from either a direct measurement of the frost thickness, or use of an appropriate correlation for frost thickness, such as in Janssen et al. (2012a, b).

Heat transfer during frost growth comprises latent and sensible heat. Heat flux generally correlates to overall mass transfer on the test surface. Figure 5.10 shows surface heat flux and temperature during frost growth for run where $T_{ch} = 0.0\ °C$, $T_{dp} = -8.3\ °C$, and $T_{ts} = -19.2\ °C$ (Experiment No. 12, Table 5.2). When humidity is injected into the chamber, a corresponding increase in the heat flux is seen at surface. The increase in heat flux is due to latent heat transfer as water molecules in the air freeze on the surface as ice crystals. The greatest rate of increase in the heat flux is seen early in the frost growth process where mass transfer is the highest. As the frost formation process continues, a reduction in both the heat and mass transfer are measured. The final heat flux, which is the sensible heat transfer, is similar to that measured for a dry (unfrosted) surface at the same conditions.

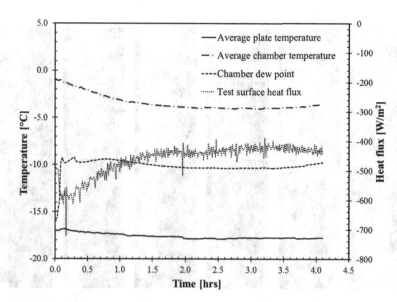

Fig. 5.10 Heat flux during frost growth (Experiment No. 12, Table 5.2)

5.3   Defrost

Figure 5.11 shows in plane and normal images of a typical defrost process. The images clearly show the changes in structure of the frost layer through the melting process. The field of view for the in plane images is 3.0 mm × 6.5 mm, and for the normal images, 3.0 mm × 2.0 mm. In plane are shown for at intervals of 4.7 s, and normal images, 5.0 s.

Minimal changes in frost structure are seen in the first few images at the initiation of the defrost process (Stage I). By the third image at ~15 s, the frost-air interface is seen moving toward the surface (Stage II), and the melt liquid is absorbed by the frost layer. By the fifth image water permeates the crystal structure. At this point, there is a steady progression of the frost-air front toward the surface. The shape of the interface does not change, and this characteristic indicates that the frost is melting at the surface. By the seventh frame at ~35 s, a visible deformation of the frost surface is seen, and all the voids in the frost layer are saturated. From this point onward, the bottom of the image shows that there is a gravity effect of melt water draining from the surface, and the existence of a suction force pulling the frost layer toward the surface. At the end of the melting process, ice crystals are visible in the water as seen in the normal images. For this frost layer, bulk downward movement of the frost layer (sloughing) off the surface is not seen, and it is inferred that surface tension is holding the water and frost to the solid surface. In the normal images, it is seen that tip crystals generally do not deform until fully saturated and remain in place on water droplets. Water drained from the surface sheds quickly, and water droplets stay in place until evaporated (Stage III).

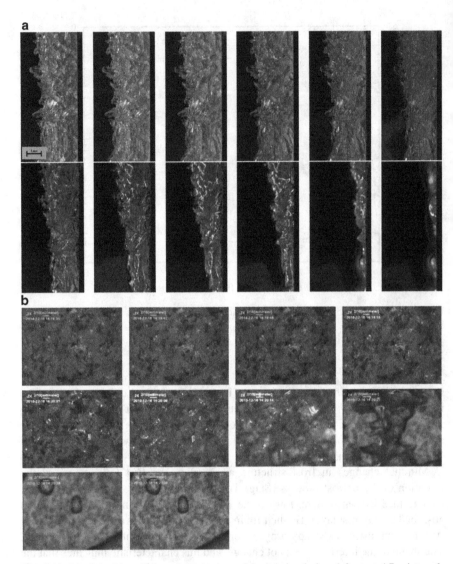

Fig. 5.11 In plane (**a**) and normal (**b**) images of frost layer during defrost at 4.7 s intervals (Experiment No. 3, Table 5.3)

From these images, it is possible to measure structural changes in the frost layer, which can be used to determine the mass transfer during the defrost process. Time-lapse images also provide a basis of reference to compare to the temperature and heat flux measurements.

From a time series plot alone, it is difficult to determine the transitions between the different stages of the defrost process. The heat flux plotted against the temperature differential with respect to the melt temperature at the surface appears to be a better indicator of the defrost process. Melting begins when the temperature difference approaches zero, and a dramatic increase in heat flux at the surface is measured, such as would occur if an infinite heat sink has been applied to the surface. The transition between the melting (Stage II) and dry-out (Stage III) processes is less defined but appears to occur when the surface temperature ~2.5 °C above the melt temperature. The change in temperature indicates a short melt period, but the heat flux indicates a longer melting stage. The video images support the longer melt period.

5.3.1   Stage I Defrost

The apparatus used to obtain our data set is capable of producing measurements of the dew point during defrost experiments, and as seen in Fig. 5.12 the dew point varies minimally. Based on a mass balance for the chamber, it is possible to calculate the mass of the water vapor in the test chamber at any given time. For the data shown in Fig. 5.12, the initial frost mass is 1.238 mg and the final mass 1.192 mg, for a total mass change of 0.046 mg in the frost layer. By comparison, the total mass of the frost layer is ~820 mg. Thus ~0.01 % of the mass is transferred by sublimation, and the mass flux is 0.51 mg/m^2 s. Table 5.3 shows the mass flux results for the present data base, and all values are on the order of milligrams per second per square meter.

Also shown in Table 5.3, are the Stefan and Lewis numbers for Stage I defrost. Recall from Chap. 3 that the reciprocal of the product of the Stefan and Lewis numbers, Γ_1, is a key parameter in the mass and energy conservation equations. A

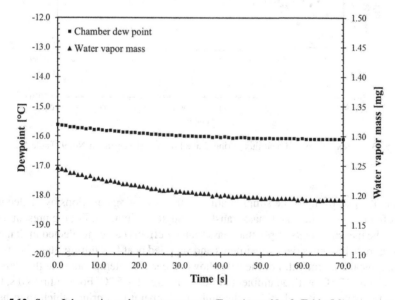

Fig. 5.12 Stage I dew point and water vapor mass (Experiment No. 9, Table 5.2)

Table 5.3 Specific mass transfer rates for Stage I defrost

Experiment	Mass transfer (mg/m² s)	St_1	Le_1	Γ_1
1	5.01	0.0155	5885	0.0110
2	1.39	0.0077	6544	0.0197
3	2.04	0.0147	12,074	0.0056
4	0.38	0.0149	8459	0.0080
5	1.61	0.0074	8485	0.0159
6	0.10	0.0149	7091	0.0094
7	0.27	0.0145	8052	0.0086
8	1.34	0.0140	6453	0.0110
9	0.51	0.0144	9065	0.0077
10	0.15	0.0135	6178	0.0120
11	0.11	0.0144	11,462	0.0061
12	0.32	0.0135	10,440	0.0071

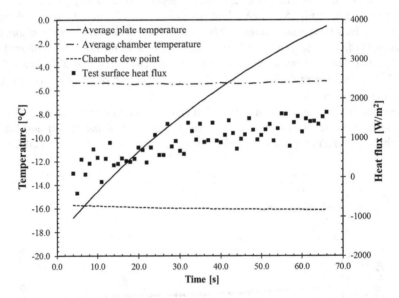

Fig. 5.13 Temperatures and heat flux during Stage I defrost (Experiment No. 9, Table 5.2)

value of $\Gamma_1 \ll 1$ allows the simplification of the governing equations by neglecting the effect of latent heat and mass transfer through sublimation. For the present data base, the measurements imply that mass transfer effects can be neglected in Stage I.

Figure 5.13 shows the temperatures and supplied heat flux during Stage I defrost for a typical experiment. For the case shown, the surface temperature at the start of defrost is −17 °C, with an ambient air temperature of −5 °C. For the first 40 s, the plate temperature is well below the ambient temperature during which time heat

Table 5.4 Energy transfer in Stage I defrost

Experiment	Input energy (J)	Heat stored (J)	Duration (s)	Average heat flux (W/m^2)	Sublimation heat flux (W/m^2)
1	108.1	9.2	36	1852	1.30E−02
2	51.8	7.7	113	285	3.63E−03
3	133.7	78.6	157	533	5.30E−03
4	121.2	45.4	159	476	9.79E−04
5	67.6	8.3	44	970	4.20E−03
6	99.0	43.9	74	836	2.49E−04
7	90.2	51.3	70	805	7.16E−04
8	101.8	37.4	65	978	3.50E−03
9	83.9	26.6	62	846	1.33E−03
10	75.4	12.0	30	1570	3.87E−04
11	75.4	45.2	39	1209	2.96E−04
12	61.5	42.1	43	893	8.38E−04

will be absorbed by the frost layer from the ambient air. Heat loss during this stage occurs only when the frost surface temperature is above the local ambient temperature, which takes place in the last 20 s of the stage. The closer the air temperature is to the melt temperature, a beneficial heating from the air will occur, and less of the heat supplied to the surface will be lost to the surrounding air. This description compares favorably to the observations of other researchers who find substantially better defrost efficiency as the ambient temperature approaches the melt temperature.

Input energy, stored energy, duration, average heat flux, and sublimation heat flux for Stage I defrost are shown in Table 5.4. Recall from Chap. 3, the initial stage of the defrost process is dominated by the sensible heating. Sensible energy storage during Stage I is,

$$E_{d1} = m_f \, c_i \left(T_m - T_{ts}\big|_{t=0} \right), \tag{5.3}$$

where T_m is the melt temperature, and $T_{ts,t=0}$ is the surface temperature evaluated at the start of the defrost process. The elapsed time for Stage I defrost is taken as the value from the moment heat is applied to the surface to the point when the surface reaches the melt temperature. By integrating the heat flux with respect to time, the input energy is calculated. Dividing the input energy by the defrost time and test surface area calculates the average heat flux, and with the mass transfer (Table 5.3), it is possible to estimate the heat transferred by sublimation. Heat transfer by sublimation is insignificant in the context of the system mass and energy balances. With the computed heat transfer, it is possible to determine the efficiency of Stage I defrost. Defrost efficiency is defined as the minimum heat input divided by the total heat input (1.2). Results are shown in Table 5.5 where a wide range is seen for the current data set, 16.3–67.0 %. When compared to all of the factors, e.g., input heat flux, degree of sensible heating, chamber ambient temperature, frost thickness, frost

Table 5.5 Defrost efficiency for Stage I

Experiment	Porosity (–)	Average chamber temperature (°C)	Sensible temperature change (°C)	Defrost efficiency (%)
1	0.54	−8.5	7.16	23.9
2	0.58	−1.2	9.18	16.3
3	0.42	0.0	19.32	58.4
4	0.48	−7.7	18.38	39.5
5	0.54	−0.6	9.46	18.1
6	0.51	−6.9	20.00	43.4
7	0.49	−5.5	19.26	55.6
8	0.53	−5.0	18.27	36.9
9	0.47	−8.2	16.63	36.2
10	0.54	−4.6	14.51	17.9
11	0.43	−4.6	19.13	58.9
12	0.45	−3.4	18.00	67.0

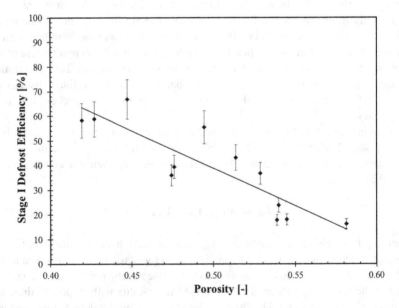

Fig. 5.14 Stage I defrost efficiency

mass and bulk porosity, defrost efficiency correlates most strongly with frost porosity (Fig. 5.14). A relatively small change in porosity produces a significant change in defrost efficiency. A less defined correlation with ambient temperature is also seen, and higher ambient temperatures produce a higher efficiency than observed by Muehlbauer (2006) and Donnellan (2007).

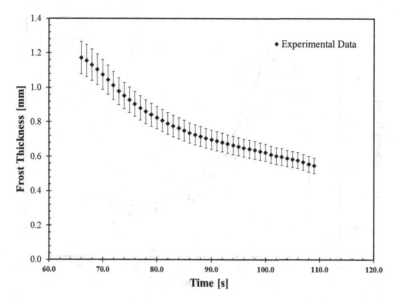

Fig. 5.15 Change in frost thickness (Experiment No. 9, Table 5.2)

5.3.2 Stage II Defrost

Stage II defrost is dominated by melting and liquid permeation. As seen in Fig. 5.11, the frost front position (thickness) changes during defrost. Figure 5.15 shows measured thickness for a typical experiment. Once the temperature at the surface reaches the melt temperature, the thickness decreases as melting occurs at the surface. A higher rate of change in the front position is observed during the first 10 s of the melting process, followed by a lower rate of change for the remaining melt process. Numerical differentiation of the interface position with respect to time produces an estimate of the front velocity. Figure 5.16 shows the estimated velocity and the predicted velocity versus time. A higher velocity is seen during the initial portion of the melt period, followed by a lower velocity until all of the frost is completely melted. The larger initial front velocity is due to the direct contact of the frost crystals with the heated surface, which aids in heat transport from the surface to the ice. As the melt progresses heat is transported into the frost layer due to permeation of the melt liquid. As the frost layer becomes fully saturated, a liquid layer forms at the surface, and the front velocity decreases as heat has to conduct through the liquid film. The good agreement between the measured and modeled front velocity shows that the melting rate is primarily influenced by the frost porosity and supplied heat flux.

Figure 5.17 shows the time-averaged front velocity for the present investigation (Table 5.2). The time-averaged velocity is defined as the change in front location from the beginning and end of the stage divided by the elapsed time. Average front velocity is strongly influenced by the magnitude of heat flux at the surface, with a

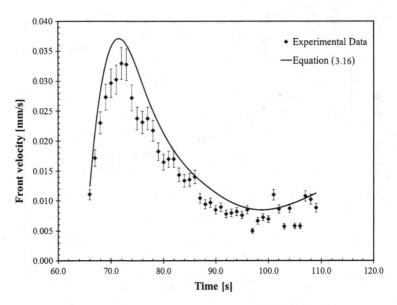

Fig. 5.16   Measured and modeled front velocity (Experiment No. 9, Table 5.2)

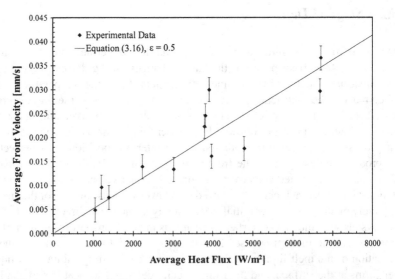

Fig. 5.17   Average front velocity versus average heat flux. Measurement compare to prediction

higher supplied heat flux resulting in a faster front velocity. For comparison, average front velocity for $\varepsilon = 0.5$ is compared to the experimental results. The limited range of porosities for the data base prevents a complete analysis of the effect of porosity on melt rate, but the strong influence of the supplied heat flux is evident.

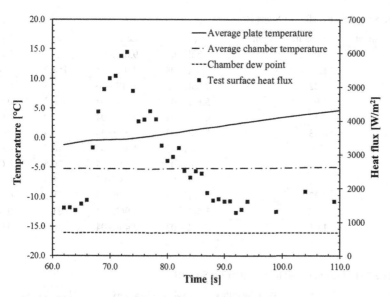

Fig. 5.18 Temperatures and heat flux during Stage II (Experiment No. 9, Table 5.2)

Similar to Stage I, a small portion of water will escape from the surface as water vapor. With an overall chamber mass balance, the mass flux is obtained (Table 5.6), and it is seen that vapor mass transfer is exceedingly small and would have an insignificant effect on overall the mass transfer during this stage. Also in Table 5.6 are the Stefan and Lewis numbers for this stage. When compared to Stage I, it is observed that the vapor Stefan number, St_{2v}, is smaller, while the Lewis number is larger, but the parameter Γ_2 is about the same order of magnitude as the Γ_1 parameter of Stage I. A value of $\Gamma_2 \ll 1$ allows simplification of the governing equations to neglect the effect of latent heat and mass transfer by sublimation.

Figure 5.18 shows the heat flux, surface temperature, dew point, and chamber temperatures for Stage II defrost. At the beginning of this stage, surface temperature is nearly constant at 0 °C for ~8 to 9 s, after which the temperature rises at a nearly constant rate. The length of the dwell time is proportional to the amount of frost in contact with the surface and the supplied heat flux. As the thin film of water is formed at the surface, heat is conducted into the water film and then into the frost layer. The ice crystals of the frost layer do not maintain contact with the solid surface.

Table 5.7 shows the average and peak heat flux measured during Stage II. Using the sublimation mass flux of Table 5.6, it possible to calculate the sublimation heat flux, and the rate of heat transfer due to sublimation is found to be insignificant when compared to the overall heat transfer rate, accounting for <0.5 % of the total heat transfer. As in Stage I, heat transfer via sublimation is negligible compared to the overall energy transfer, and heat transfer at the surface is essentially absorbed by the melting frost layer.

Table 5.6 Mass transfer for Stage II defrost

Experiment	Mass flux (mg/m² s)	St_{2v}	Le_{2v}	Le_{2f}	Γ_2
1	7.08	6.50E−03	9.68E+03	28.9	0.016
2	1.32	9.29E−04	4.03E+04	158.5	0.027
3	2.22	7.66E−05	1.20E+06	48.3	0.011
4	2.94	5.89E−03	1.47E+04	46.2	0.012
5	6.32	4.65E−04	9.93E+04	71.9	0.022
6	0.43	5.28E−03	1.34E+04	14.4	0.014
7	0.79	4.21E−03	1.80E+04	16.0	0.013
8	1.28	3.83E−03	1.54E+04	26.5	0.017
9	0.01	6.28E−03	1.48E+04	35.4	0.011
10	2.27	3.53E−03	1.55E+04	55.2	0.018
11	2.46	3.52E−03	2.96E+04	21.1	0.010
12	1.61	2.61E−03	3.40E+04	15.5	0.011

Table 5.7 Defrost time and heat transfer for Stage II

Experiment	Defrost time (s)	Peak heat flux (W/m²)	Average heat flux (W/m²)	Sublimation heat flux (W/m²)
1	26	6456	3792	9.97
2	91	2663	1042	3.42
3	152	3227	1389	5.75
4	90	2898	1203	7.62
5	30	5452	3015	8.91
6	30	6400	3914	1.12
7	41	7036	3818	2.05
8	55	8093	3969	3.31
9	26	6031	3432	3.98
10	21	9437	4788	5.89
11	23	13,051	6676	6.39
12	20	9773	6699	4.19

The duration of Stage II is difficult to determine from temperature data. It appears that this stage is very brief across our data set, but the heat transfer data imply a longer second stage. Digital photographs agree with the longer length of time for the stage. Table 5.7 shows the length of second stage defrost time as measured by the heat flux data. Recall that the frost front velocity is proportional to the frost density, surface heat flux, and latent heat of fusion. Thus, Stage II melt time can be estimated by,

$$\Delta t \approx \frac{\lambda_{if}\rho_{f,o}}{q''_{ts}}\Delta\delta. \tag{5.4}$$

Figure 5.19 shows calculated and measured defrost time for Stage II, and the agreement is fair to good considering the assumptions underlying the calculations.

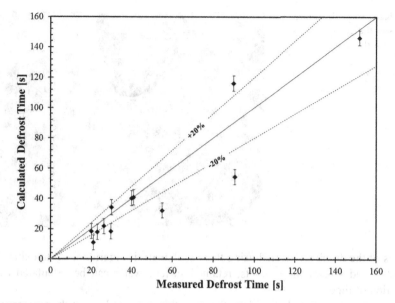

Fig. 5.19 Temperatures and heat flux during Stage II (all experiments)

The simple model provides an acceptable estimation of the defrost time, with most of the calculated times falling within a ±20 % tolerance band.

The defrost efficiency for Stage II is,

$$\eta_{d,2} = \frac{\lambda m_f}{E_{d,2}} \tag{5.5}$$

where m_f is the mass of the frost, λ is the latent heat of fusion, and E_{d2} is the supplied energy The measured defrost efficiency is on the order of 95–98 %, which is the largest of the three stages. The reasons for the high defrost efficiency are that the temperature difference between the air and frost surface is small, limiting the heat transfer rate, and the stage is brief, limiting the amount of time for heat to escape.

5.3.3 Stage III Defrost

Draining and evaporation of the melt liquid dominates the heat and mass transfer processes in Stage III defrost. From the visual record, film flow during draining is a relatively quick process, taking just a few seconds. At the conclusion of the draining process, the surface is wetted with stationary water droplets. Mass transfer through evaporation is determined by measuring the change in volume of the droplets of the surface. Figure 5.20 shows a typical an image from the dry out process with the

Fig. 5.20 Stage III surface drops with digitally determined edges

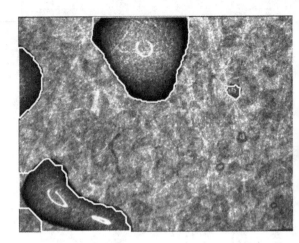

edges of the droplets detected using the visual analysis algorithm. By tracking the number and size of the droplets, retained surface water can be calculated as a function of time.

In terms of estimated mass data, the mass flux of water leaving the surface is,

$$m''_{v,w} = m''_{v,3}\frac{A_s}{A_w} = m''_{v,3}\frac{1}{f}, \qquad (5.6)$$

where f is the ratio of the wetted to surface area, A_w/A_s, and the droplet volume and wetted area are expressed by the characteristic diameter. The change in wetted area is proportional to the change in mass, and mass transfer occurs across the wetted surface area. Mass transfer from the wetted surface is determined from the whole surface mass transfer coefficient. Thus, the time dependent change in wetted area fraction is,

$$f = f_0\left(1 - \frac{h_m\Delta\rho_v}{m_0/A_s}t\right), \qquad (5.7)$$

where, f_0 and m_0 are the initial wetted area fraction and retained droplet mass, both functions of surface wettability. Once the mass transfer from a drop is known, the mass transfer coefficient can be estimated. Table 5.8 lists measured mass transfer and mass transfer coefficients for our data set. By comparison, the mass transfer coefficient estimated by the heat and mass transfer analogy is ~0.045–0.060 m/s, and these values are in good agreement with the measurements made during this study.

In Fig. 5.21, Eq. (5.7) is plotted with experimental results for two cases. The model is in reasonable agreement to the measured data. The rate of change of the retained mass is dependent on a number of factors, but the primary factor affecting evaporation is the water vapor pressure difference between the surface and the ambient air. As expected, a higher vapor pressure difference results in a greater rate of change and shorter dry-out time. The duration of Stage III is significantly longer

Table 5.8 Mass transfer for Stage III defrost

Experiment	Mass transfer rate (mg/m² s)	h_m (m/s)
1	0.56	0.051
2	0.26	0.030
3	0.50	0.055
4	0.41	0.039
5	0.61	0.064
6	0.55	0.052
7	0.50	0.047
8	0.56	0.051
9	0.51	0.048
10	0.94	0.083
11	0.98	0.083
12	0.72	0.070

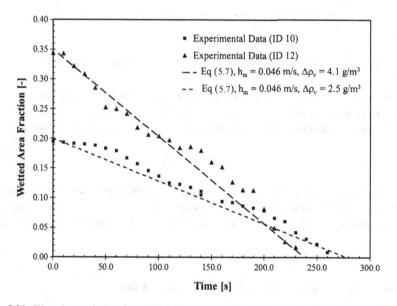

Fig. 5.21 Wetted area during Stage III defrost compared to prediction. *Filled rectangle* Experiment No. 10. *Filled triangle* Experiment No. 12. (Table 5.2)

than that of the other two stages, ~200 to 300 s, owing to the slower nature of the evaporation process.

Figure 5.22 shows typical the temperatures and surface heat flux for Stage III. From the temperature plot it is impossible to determine the exact transition from the second to third stage when all of the frost crystals have melted. With the visual record however, it appears that all of the ice crystals melt ~120 s into this experiment. Over the course of the dryout, heat flux remains approximately constant, while the surface temperature increases at an almost constant rate.

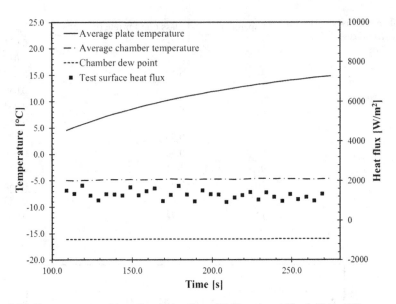

Fig. 5.22 Temperatures and heat flux during Stage III (Experiment No. 9, Table 5.2)

As dryout proceeds, the water droplets decrease in size and expose more of the dry surface to the ambient air. Thus the total heat transfer from the surface will be a combination of the latent heat of evaporation and sensible heat from the dry surface,

$$Q_3 = Q_{3,\mathrm{sn}} + Q_{3,\mathrm{lt}}. \tag{5.8}$$

Assuming that the water droplets are at the same temperature as the surrounding surface, sensible heat exchange occurs over the entire surface, A_s, while latent heat transfer is occurs over the wetted area, A_w. Thus, the heat flux from the surface is,

$$q_3'' = q_{3,\mathrm{sn}}'' + fq_{3,\mathrm{lt}}''. \tag{5.9}$$

A total surface heat transfer coefficient, h_3^* can be defined,

$$h_3^* = h_{3,\mathrm{sn}} + fh_{3,\mathrm{lt}}, \tag{5.10}$$

where $h_{3,\mathrm{sn}}$ and $h_{3,\mathrm{lt}}$ are the heat transfer coefficient for sensible latent heat exchange respectively. The latent heat transfer coefficient can be expressed,

$$h_{3,\mathrm{lt}} = \frac{q_{3,\mathrm{lt}}''}{\Delta T} = \frac{\lambda_{\mathrm{fg}} m_{v,3}''}{\Delta T} = \frac{\lambda_{\mathrm{fg}} h_m \Delta \rho_v}{\Delta T}, \tag{5.11}$$

where, $\Delta \rho_v$ and ΔT are the vapor and temperature potentials between the surface and ambient air. Combination of (5.7), (5.10) and (5.11) yields the total surface heat transfer coefficient,

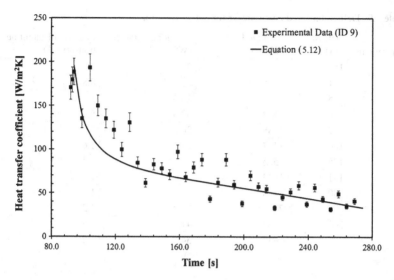

Fig. 5.23 Measured and predicted heat transfer coefficients for Stage III (Experiment No. 9, Table 5.2)

$$h_3^* = h_{3,sn} + f_0\left(1 - \frac{h_m \Delta\rho_v}{m_0/A_s}t\right)h_m\lambda_{fg}\frac{\Delta\rho_v}{\Delta T}, \tag{5.12}$$

From (5.12) it is seen that the total surface heat transfer coefficient is related to a sensible heat transfer coefficient, and a time-dependent latent heat transfer coefficient. Furthermore, the latent heat transfer coefficient decays at a quadratic rate. Figure 5.23 shows a comparison of measured heat transfer coefficients for a typical dry out to (5.12) with $h_{3,sn} = 40$ W/m² K, $f_0 = 0.45$, $m_0 = 0.05$ kg/m² and $h_m = 0.48$ m/s. Good agreement is seen between measured and calculated heat transfer coefficients.

Table 5.9 summarizes the defrost time and heat transfer rates to raise the test surface to 20 °C, which is at a temperature where most of the melt liquid is evaporated from the surface.

An observation is that even with high supplied heat fluxes most of the heat is transported through sensible heat exchange, with only 2–10 % of it transported through sensible heat transfer.

Defrost efficiency in Stage III is inherently the lowest of the three stages owing to the evaporation mechanism. In addition to the latent heat transfer, there is significant sensible heat loss to the ambient air, which in a refrigerated container would be the conditioned air space. In practice prolonged heating of the heat exchanger mass requires a greater amount of cooling to reduce the coil temperature during defrost recovery. Rearranging (5.7), the length of the defrost time is,

$$t = \left(1 - \frac{f}{f_0}\right)\frac{m_0}{h_m\Delta\rho_v}, \tag{5.13}$$

Table 5.9 Defrost time and heat transfer for Stage III defrost

Experiment	Defrost time (s)	Average heat flux (W/m^2)	Latent heat transfer (W/m^2)	Fraction of latent heat transfer (%)
1	153	1755	34.3	2.0
2	101	245	21.7	8.8
3	125	1469	26.6	1.8
4	178	532	32.8	6.2
5	77	1388	20.2	1.5
6	166	1467	24.1	1.6
7	175	535	28.7	5.4
8	181	1643	29.1	1.8
9	180	1338	18.1	1.4
10	75	2492	30.8	1.2
11	201	2847	41.9	1.5
12	195	2837	34.0	1.2

In practice, the intent is to minimize defrost time and maximize defrost efficiency, and thus decreasing the initial mass of the retained melt liquid will have the largest impact. The only way to affect the retained mass is to control the surface wettability. Another way to decrease the defrost time is to leave residual moisture on the surface, which is a common practice in the industry. The retained liquid refreezes during the subsequent cooling cycle and generally does not have a negative effect on system performance other than the frozen droplets become as a nucleation site for subsequent frosting cycles. This may lead to more frequent defrost cycles.

5.4 Summary

Table 5.10 summarizes overall the defrost efficiency for Stages I–III combined for the various parameters of our data set. Defrost efficiency is defined as the minimum energy to melt the frost layer divided by the total heat input (1.2). Defrost is taken to be complete when the surface temperature reaches 20 °C. Defrost efficiencies range from 25.6 to 93.7 %. Of all of the factors measured (mass, porosity, thickness, etc.), initial frost thickness yields the strongest correlation to overall efficiency, and efficiency increases with thicker frost layers (Fig. 5.24).

An optimal frost thickness cannot be determined with the present data base because there is insufficient data to determine if defrost efficiency decreases for exceedingly thick frost layers. However it appears that thicker frost layers yield better defrost efficiency owing to the insulating properties of the frost itself. For the thick layer, a substantial temperature gradient is formed within it when surface heating is applied. While the crystals near the surface are melting, there is ice at the frost-air interface which limits heat loss. A large portion of the applied energy at the

Table 5.10 Overall defrost efficiency for Stages I–III

Experiment	Mass (g)	Melt energy (J)	Energy input (J)	Defrost efficiency (%)
1	0.68	247	721	34.3
2	0.43	149	359	41.6
3	2.13	775	1012	76.6
4	1.31	471	598	78.8
5	0.45	155	499	31.1
6	1.16	420	448	93.7
7	1.41	506	605	83.6
8	1.08	387	630	61.5
9	0.84	303	569	53.4
10	0.43	155	607	25.6
11	1.25	449	570	78.9
12	1.19	441	562	78.4

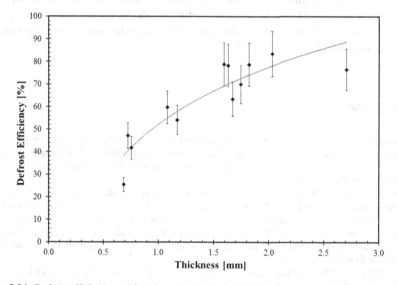

Fig. 5.24 Defrost efficiency as a function of initial frost thickness at the start of defrost. Error bars are estimated r.m.s. uncertainty

surface is absorbed by either a thick frost layer or a frost layer with low porosity (high bulk density). Frost layers with low porosity appear to absorb more of the surface heat transfer and thus have lower heat loss. Heat and mass transfer through sublimation during this stage (Stage I) are insignificant. For thin layers, a more uniform temperature layer results in a greater losses to the ambient air. This conclusion is supported by the visual record: For melting of thick frost layers, melt liquid is seen at the surface, while ice crystals at the frost-air interface remain solid for several seconds over a fairly large fraction of our total defrost time for a small test surface of 38 mm × 38 mm. For thin layers, the entire frost layer appears to melt instantaneously. Based on our results, the optimal defrost efficiency occurs for frost thickness of ~1.5 to 2.0 mm.

The primary factor effecting Stage I defrost is the bulk porosity at the start of the process. A large portion of the applied surface heat is absorbed by the frost layer. Frost layers with low porosity (high density) absorb more of the heat, and thus minimize heat loss through it. Heat and mass transfer through sublimation during this stage is found to be insignificant. Stage II defrost is dominated by melting. The frost front velocity is determined via digital analysis of the in plane images and is found to vary with supplied heat and porosity. A higher heat transfer rat results in faster melt velocity and thus shortened defrost times. Low frost porosity has the effect of increasing the defrost time. Effects of sublimation are negligible. Defrost efficiency for this stage is nearly 100 % with little heat lost to the surroundings. Evaporation of the melt liquid dominates Stage III defrost. This stage has the lowest defrost efficiency because most of the surface heating is lost through sensible heat exchange with the ambient air. A heat transfer model for the wetted surface captures both sensible and latent heat exchange effects. Latent heat exchange is correlated to an area reduction of the water droplets, which is expressed by a mass transfer coefficient.

References

Donnellan W (2007) Investigation and optimization of demand defrost strategies for transport refrigeration systems. Doctoral dissertation, Galway-Mayo Institute of Technology, Galway

Janssen DD (2011) Experimental strategies for frost analysis. Master's thesis, University of Minnesota, Minneapolis

Janssen DD, Mohs WF, Kulacki FA (2012a) Modeling frost growth—a physical approach. In: Proceedings of the 2012 ASME summer heat transfer conference, paper no. HT2012-58054

Janssen DD, Mohs WF, Kulacki FA (2012b) High resolution imaging of frost melting. In: Proceedings of the 2012 ASME summer heat transfer conference, paper no. HT2012-58061

Lee YB, Ro ST (2002) Frost formation on a vertical plate in simultaneously developing flow. Exp Therm Fluid Sci 26:939–945

Mohs WF (2013) Heat and mass transfer during the melting process of a porous frost layer on a vertical surface. Doctoral dissertation, Universality of Minnesota, Minneapolis

Muehlbauer J (2006) Investigation of performance degradation of evaporators for low temperature refrigeration applications. Master's thesis, University of Maryland, College Park

Na B (2003) Analysis of frost formation in an evaporator. Doctoral dissertation, Pennsylvania State University, University Park

Chapter 6
Solution of Defrost Model

Abstract The differential equations describing heat and mass transfer during each stage of defrost are solved, and results are compared to measurement. Numerical solutions for defrost Stages I and II are obtained, while an analytical solution for Stage III is possible. For Stage I, the duration of the stage is predicted and validated by experiment. For Stage II, melt front and duration of the stage are predicted and in good agreement with measurement. For Stage III, the solution under predicts defrost time but generally simulating the general trends for duration seen in the experiments. We conclude with a graphical summary that indicates that further research is needed for measurements with greater super saturation and subcooling.

Keywords Duration of defrost • Parameter variation • Comparison of experiment and theory

Nomenclature

Bi	Biot number, $h\delta/k_f$
C_1, C_2	Constants (6.6)
h	Heat transfer coefficient (W/m^2 K)
k	Thermal conductivity (W/m K)
Le_3	Lewis number for Stage III
Q	Heat transfer (W)
S	Water content (–)
St_3	Stephan number for Stage III (3.73)
t	Time (s)
u	Convective velocity of the melt stream (3.16)
U	Dimensionless velocity of frost thickness in Stage II (–)

© Springer International Publishing Switzerland 2015
W.F. Mohs, F.A. Kulacki, *Heat and Mass Transfer in the Melting of Frost*,
SpringerBriefs in Applied Sciences and Technology,
DOI 10.1007/978-3-319-20508-3_6

Greek Symbols

δ Frost thickness (m)

ε Porosity (–)

Γ_1 Inverse of Lewis and Stephan number product for Stage I, 1/LeSt (3.39)

Γ_3 Inverse of Lewis and Stephan number product for State III (3.73)

η Dimensionless thickness, y/δ_0

η_d Efficiency (6.1)

η_i Dimensionless frost thickness in Stage II (6.3)

λ_{if} Latent heat of fusion (J/kg)

θ Dimensionless temperature $(T - T_s)/(T_m - T_s)$

τ Dimensionless time, $\alpha_f t/\delta_0^2$

Subscripts

0   Initial time

1,2,3   Denoting the defrost stage

a   Ambient

c   Critical value

f   Frost

fs   Frost surface

s   Surface

w   Water

Superscripts

m   Integration step

6.1   Stage I Diffusion

The fundamental equations for Stage I defrost have been developed in Chap. 3. Figure 6.1 shows the change in dimensionless temperature within the frost layer for various dimensionless times and selected values of the Biot numbers and Γ_1. A relatively large value of Γ_1 is chosen here, but our experiments yield a much smaller value. The simulation starts with a steady state temperature distribution that would exist at end of the frosting process and ends when the surface temperature reaches T_m, or $\theta = 1.0$, which is the point where melting would begin at the surface. A significant change in shape of the temperature distribution is seen early in the heating process. Here most of the heat applied is absorbed by the frost layer. As time progresses, the temperature profile approaches a linear steady state.

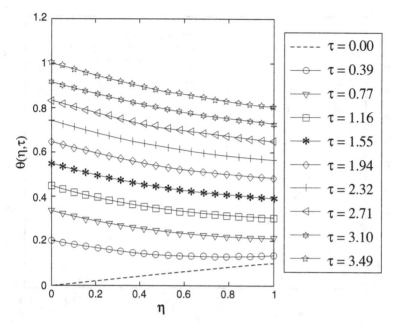

Fig. 6.1 Stage I frost temperature distribution. $\Gamma_1 = 0.01$, $\mathrm{Bi_s} = 0.2$, $\mathrm{Bi_{fs}} = 0.1$, $\theta_{fs,0} = 0.1$, $\theta_a = 0.75$

The two factors that have the greatest effect on Stage I duration are porosity and the boundary condition at the surface. Figure 6.2 shows the effects of these factors on defrost time. Recall that as porosity increases, there is a greater potential of heat transport due to sublimation. The greater heat transport reduces the time it takes for the surface to reach the melt temperature. This effect is most pronounced at large values of porosity. For most frost layers, $0.4 < \varepsilon < 0.6$ (volume averaged), and the effect heat transfer by sublimation is minimal which is borne out by our measurements. Thus changes at the surface are found to have a greater effect on the duration of Stage I. Increasing $\mathrm{Bi_s}$, such as raising the heat flux at the surface, can dramatically reduce the time.

Table 6.1 summarizes the inputs and results for the simulations that are compared to the experiments. The effect of Γ_1 is included in the simulation, but it has a minimal effect on the results. Our model generally under predicts defrost time, but follows the trend of higher surface Biot number resulting in a shorter duration. The disagreement is due to a number of factors. The model assumes one-dimensional heat transfer, while in the experiments two-dimensional effects are observed. The model also assumes a constant heat flux at the surface, while there was difficulty in maintaining a constant heat flux at the surface during the defrost experiments.

Using the results of the simulation, it is possible to calculate the defrost efficiency of the first stage. Recall, the first stage defrost efficiency is defined,

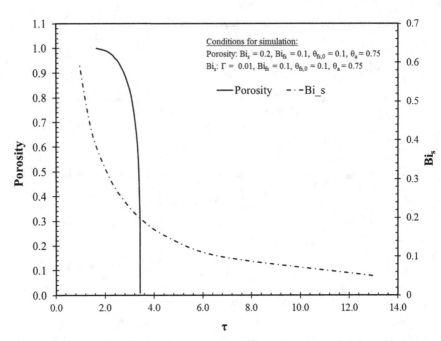

Fig. 6.2 Calculated effects of porosity and surface boundary condition on Stage I defrost time. For porosity: $Bi_s = 0.2$, $Bi_{fs} = 0.1$, $\theta_{fs} = 0.1$, $\theta_a = 0.75$. For Bi_s: $\Gamma = 0.01$, $Bi_{fs} = 0.1$, $\theta_a = 0.75$. $\varepsilon = (\rho_i - \rho_f)/(\rho_i - \rho_a)$. $\theta_a = (T_a - T_s)/(T_m - T_s)$. $Bi_s = \delta_f q_s''/k_f (T_m - T_s)$. $Bi_{fs} = \delta_f q_s''/k_f (T_a - T_{fs})$. $\tau = \alpha_f t/\delta_f^2$

Table 6.1 Summary of simulation results for Stage I

Experiment	Γ_1	Bi_s	Bi_{fs}	$\theta_{fs,0}$	θ_a	τ Experiment	τ Model
1	0.01	0.28	0.07	0.10	0.58	11.5	2.7
2	0.02	0.07	0.06	0.11	0.88	38.3	10.6
3	0.01	0.14	0.12	0.13	1.00	9.4	4.8
4	0.01	0.10	0.09	0.05	0.60	19.4	8.3
5	0.02	0.21	0.05	0.08	0.94	30.4	3.6
6	0.01	0.19	0.10	0.07	0.65	9.3	4.1
7	0.01	0.21	0.11	0.10	0.71	6.8	3.5
8	0.01	0.24	0.10	0.11	0.73	8.8	3.0
9	0.01	0.12	0.06	0.04	0.56	18.5	7.1
10	0.01	0.17	0.04	0.05	0.74	23.3	4.8
11	0.01	0.20	0.07	0.08	0.76	6.7	3.8
12	0.01	0.17	0.08	0.09	0.80	6.8	4.5

$$\eta_d = \frac{\text{Energy absorbed by frost}}{\text{Total energy input}} = \frac{\Sigma(Q_s - Q_{fs})\Delta t}{\Sigma Q_s \Delta t}. \tag{6.1}$$

With the relations defined for the Stage I Biot numbers at the surface and the frost-air interface,

Equation (6.1) can be expressed,

$$\eta_{d,1} = 1 - \frac{Bi_{fs}}{Bi_s}\sum(\theta_{fs} - \theta_a)\Delta\tau. \tag{6.2}$$

Defrost efficiency will be less than 100 % when the frost surface temperature is greater than the ambient temperature, and greater than 100 % when the ambient temperature is higher than the frost surface temperature. In the first case, surface heat will be lost from the frost and into the ambient air. In the second case, heat transfer from the ambient air will aid the defrost process. In most situations at the start of the defrost process, the temperature of the frost surface will be below the ambient air temperature resulting in a high defrost efficiency. As time progresses, the frost surface temperature will exceed the ambient temperature resulting in a loss of heat to the ambient air, and a reduction of the defrost efficiency. The temperature difference is scaled by the ratio of the Biot numbers at the frost-air and frost-solid interfaces. A higher surface Biot number results in a lower ratio, reducing the temperature effects on the defrost efficiency.

Figure 6.3 shows the effect the Biot number ratio on Stage I defrost efficiency over a range of dimensionless ambient air temperature. When the ambient

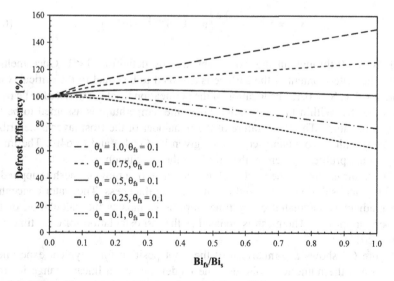

Fig. 6.3 Effect of boundary conditions on Stage I defrost efficiency

temperature is near the melt point, defrost efficiencies greater than 100 % can be achieved. Once the dimensionless temperature falls below 0.5, efficiency is less than 100 %, but can be improved by with a high heat transfer at the frost-surface interface relative to that at the frost-air interface. Ways to maximize the defrost efficiency are by selecting heat sources that can be supplied at a high rate, such as electrical resistance heating and that reduce heat transfer to the ambient air by limiting convection and thermal radiation.

6.2 Stage II Melting and Permeation

Stage II defrost is modeled as the melting of the frost at the heated surface. When the surface temperature exceeds the melt temperature, melting begins. The melt liquid is drawn into the open pores of the frost layer, and the thickness of the frost layer will decrease. The rate of change of the frost layer thickness is given by (3.16). Applying an upwind finite difference, the dimensionless position of the thickness is,

$$\eta_i^m = \eta_i^{m-1} - U\Delta\tau, \tag{6.3}$$

where the velocity ratio, U, is equal to 1 for a constant supplied heat flux.

As the melt liquid is absorbed into the frost layer, a permeation layer grows away from the surface. The growth of the permeation layer is described by (3.57)with a water content that is governed by (3.58), Applying a linear upwind approximation, the water content is,

$$S_i^m = S_i^{m-1} + \frac{\Delta\tau}{\Delta\eta}\left[1 - U\left(S_i^{m-1} - S_{i-1}^{m-1}\right)\right]. \tag{6.4}$$

Initially $S = 0$ throughout the frost layer, and by definition $S < 1$. Once melting starts, the water content at the surface is assumed to be equal to the critical water content value, $S_c = 0.1$. The location of the permeation front is estimated to be point where $S < S_c$. Within the permeation layer, the temperature is assumed to be the melt temperature. The temperature of the remainder of the frost layer is described by (3.63) with the boundary conditions given by (3.64a) and (3.64b). The initial temperature profile is given by the profile at the conclusion of Stage I.

Due to the change in the frost thickness in Stage II, a moving mesh is applied to the numerical solution, initialized for the starting thickness. The water concentration gradient is calculated every time step, as it impacts the calculation of the temperature profile. The mesh is updated as the last calculation for each time step (Mohs 2013).

Figure 6.4 shows a comparison of the front position for a typical experiment (No. 9) and the numerical solution. The model predicts a linear change in front position because a constant velocity ratio was used. During the experiment the front velocity is not constant owing to the variability of the supplied heat flux during.

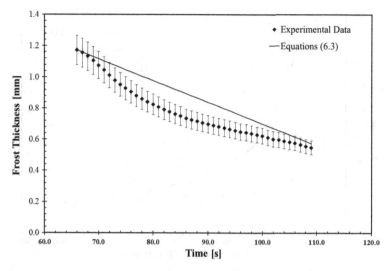

Fig. 6.4 Comparison of frost thickness measurements with model results. Experiment No. 9, Table 5.2

Table 6.2 Summary of simulation results for Stage II defrost

Experiment	u_0 (mm/s)	$Le_{2,f}$	Γ_2	$Bi_{2,s}$	$Bi_{2,fs}$	Dimensionless time	
						Experiment	Model
1	0.022	28.9	0.016	1.50	0.07	0.54	0.45
2	0.005	158.5	0.027	2.35	0.06	0.58	0.55
3	0.008	48.3	0.011	7.61	0.12	0.42	0.41
4	0.010	46.2	0.012	0.70	0.09	0.48	0.52
5	0.013	71.9	0.022	1.93	0.05	0.55	0.54
6	0.030	14.4	0.014	2.77	0.10	0.51	0.59
7	0.025	16.0	0.013	3.69	0.11	0.50	0.50
8	0.016	26.5	0.017	3.94	0.10	0.53	0.51
9	0.014	35.4	0.011	0.78	0.06	0.48	0.48
10	0.018	55.2	0.018	2.21	0.04	0.54	0.48
11	0.030	21.1	0.010	4.80	0.07	0.43	0.43
12	0.037	15.5	0.011	7.06	0.08	0.45	0.42

While a constant velocity ratio does not match the data over the whole range, the final front position is predicted with good accuracy owing to the integral nature of the phase change process, wherein most of the supplied energy is absorbed by the frost layer resulting in a nearly uniform melt rate. Using a non-uniform heat flux for the simulation would yield a closer match to the experimental results.

Table 6.2 lists numerical results using experimentally derived inputs. The model predicts the melt duration within 20 % of the measured value, with most of the data within 10 %. From the dimensionless time, it is difficult to infer the effect of

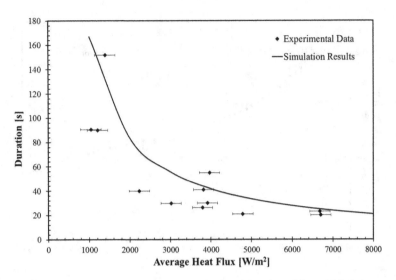

Fig. 6.5 Comparison of measured duration and predicted duration of Stage II defrost versus of averaged heat flux

boundary conditions on the defrost time. Converting the model back to engineering units a better correlation can be seen. Figure 6.5 shows a comparison of the experimental results with the numerical solution assuming a constant initial frost thickness and porosity. A strong relation between the Stage II duration and surface heat flux is confirmed. A higher heat flux results in a shorter duration of defrost, but as the heat flux is dramatically increased, the effect on shortening the duration is diminished. From the figure it appears a optimal heat flux would be ~5000 W/m^2 considering a trade-off between supplied heat and duration of the melt stage.

6.3 State III Defrost

Stage III defrost is modeled as evaporation of a thin water film on the heated surface. Heat and mass transfer are expressed by the coupled ordinary differential (3.71) and (3.72). Equation (3.72) has the solution of the form,

$$\theta = \frac{C_2}{C_1}\left(e^{C_1\tau} - 1\right), \tag{6.5}$$

where C_1 and C_2 are,

$$C_1 = \left[Bi_{3,w} + \frac{1}{St_3}\frac{1}{Le_3}\right], \quad C_2 = Bi_{3,s}. \tag{6.6}$$

The constant C_1 represents the sum of the sensible and latent heat transfer from the film surface, while C_2 represents the heat transfer into the film from the surface. Substituting (6.5) into (3.71) and integrating, the film thickness is,

$$\eta_w = 1 - \frac{C_2}{C_1^2 Le_3}\left(e^{C_1\tau} - 1 - C_1\tau\right), \tag{6.7}$$

The magnitude of the heat transfer at the surface has a large effect on the duration of Stage III defrost. Figure 6.6 shows the film thickness and the temperature during the dry out process for various supplied heat fluxes with all other factors held constant. Like the experimental results an increase in heat flux shortens the defrost time. An important observation from the simulation is the slow rate of change in the film thickness for a substantial period of the dry-out process. For example, the first 20 % of reduction in film thickness takes over 60 % of the total time. This is a result of the evaporation process, as the vapor pressure at the water surface needs to exceed the partial pressure of water in the air. Also, for evaporation to occur, the final temperature at dry out is nearly the same, regardless of the rate of heat transfer.

Figure 6.7 shows the effect of the ambient temperature when other factors are held constant. Lower air temperatures are result in a longer dry out process. The reason for the longer time is the diminished capacity of the colder air to hold moisture, slowing evaporation.

Table 6.3 summarizes the simulation results using the inputs from the experiments discussed in Chap. 5. Both measured and modeled defrost times are reported.

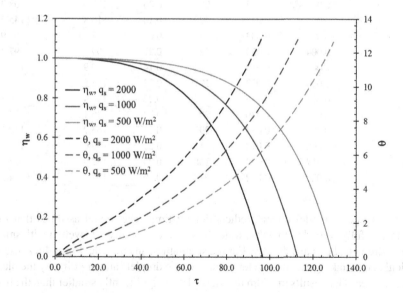

Fig. 6.6 Film thickness and temperature during Stage III for selected values of test surface heat flux. $\eta_w = y/\delta_{w,0}$. $\theta = (T_w - T_{w,0})/(T_a - T_{w,0})$. $\tau = \alpha_w t/\delta_{w,0}^2$

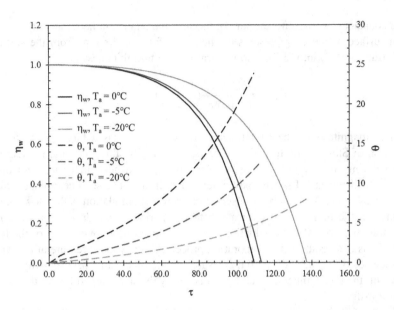

Fig. 6.7 Effect of ambient temperature on film thickness and film temperature during Stage III defrost. $\eta_w = y/\delta_{w,0}$. $\theta = (T_w - T_{w,0})/(T_a - T_{w,0})$. $\tau = \alpha_w t/\delta_{w,0}^2$

Table 6.3 Summary of simulation results for Stage III

Experiment	Bi_{3s}	Bi_{3w}	Le_3	St_3	τ Experiment	τ Model
1	0.008	0.025	294	0.26	173.6	108.3
2	0.002	0.018	470	0.10	210.8	145.2
3	0.024	0.033	334	0.09	77.37	63.9
4	0.004	0.033	226	0.22	109.7	97.3
5	0.009	0.019	562	0.14	149.9	133.7
6	0.010	0.032	195	0.22	112.5	75.6
7	0.011	0.035	184	0.22	100.4	70.8
8	0.010	0.030	204	0.23	136.1	78.6
9	0.006	0.028	297	0.27	163.7	110.1
10	0.009	0.018	311	0.22	159.6	114.2
11	0.017	0.032	168	0.25	132.9	63.6
12	0.021	0.032	212	0.20	127.5	64.3

In all cases, the model under predicts defrost time. The model assumes that the surface is fully wetted with a thin water film, while in reality it is wetted with small water droplets, i.e., partially dry. For the simulation, an equivalent film thickness is calculated by taking the total water volume of the droplets and dividing by the plate surface area. This results in a film thickness that is significantly smaller than the true height of the drops. Increasing the film thickness results in a longer defrost time. The inputs of the simulation are held constant, while in the experiment both heat transfer and the ambient temperature vary owing to difficulty in controlling them. While the model under-predicts, it demonstrates similar trends to those seen the experiments and can be useful as a tool to understand the factors effecting the defrost process.

6.4 Conclusion

The heat and mass transfer model developed in Chap. 3 has been solved and results compared to measurements for a complete defrost sequence. The model allows for detailed analysis of the defrost process, and an evaluation of parameters which can be controlled to effect defrost performance. The differential equations governing heat and mass transfer in Stages I and II are numerically solved. The governing equation for Stage III is analytically solved. Experimentally verified simplification of the equations via scale analysis has facilitated these solutions.

The scope of the experiments that comprise the data base for evaluation of our model is shown in Fig. 6.8. For Stage I, porosity has a minimal effect on the duration of the stage. The surface heat flux has a more significant effect, but a very high heat flux has a diminishing effect on reducing defrost duration. Confirming the experimental study, heat and mass transfer through sublimation was shown to be insignificant to the overall transfer processes. Defrost efficiency greater than 100 % is achievable when the ambient air temperature is near the melt temperature. This is primarily due to heat transfer from the ambient air into the frost layer, aiding the heating process. For lower ambient air temperature, defrost

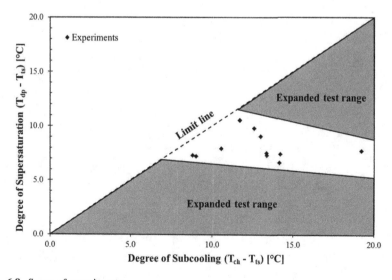

Fig. 6.8 Scope of experiments

efficiency can be favorably improved by increasing the heat transfer rate into the frost layer.

Stage II is dominated by the melting, and the model is capable of tracking the change in frost thickness and the location of the permeation front. The front velocity varies with the supplied heat and porosity. Higher heat transfer rates at the surface result in a faster melt and thus a shortened stage duration. Effects of sublimation are negligible on the overall heat and mass transfer. The current model does not include surface physics, capillary forces, and frost slumping.

An analytical solution for Stage III defrost adequately determines the effects of the heat transfer rate and ambient air temperature. This stage of defrost is dominated by evaporation of the liquid film and surface droplets. Surface physics and capillary forces are not included in the model. Similar to the earlier stages, a high heat transfer rate reduces the defrost duration. A lower ambient air temperature lengthens the duration of the melt.

Reference

Mohs WF (2013) Heat and mass transfer during the melting process of a porous frost layer on a vertical surface. Doctoral dissertation, Universality of Minnesota, Minneapolis

Index

C

Comparison of experiment and theory, 90–92

D

Defrost, 2–8, 10, 12–18, 21–41, 45, 48–54, 57–96
Defrost efficiency, 6–8, 12, 16–18, 45, 50–51, 71, 72, 77, 81–85, 87, 89, 95
Digital analysis, 45, 51, 52, 58, 63, 76, 78, 84
Dry out, 5, 16, 31–33, 40–41, 51, 58, 60, 69, 77, 78, 81, 93
Duration of defrost, 7, 8, 14, 16–18, 49, 71, 76, 78, 87, 91–93, 95, 96

F

Frost formation, 9–12, 46, 60, 66
Frost growth, 2, 7, 9–12, 17, 18, 46, 51, 52, 54, 59–67
Frost layer, 2, 4–6, 10, 12–16, 18, 23–31, 33, 36, 38, 39, 45, 49–52, 60, 61, 63, 66–69, 71, 73, 75, 82–84, 86, 87, 90, 91, 96
Frost profile, 29, 51, 62
Frost properties, 2, 9–10, 18, 23, 36, 37, 48, 58
Frost thickness, 2, 10, 12, 14, 15, 18, 45, 52–53, 59, 62, 63, 66, 71, 73, 82–84, 90, 91, 96

H

Heat transfer, 2, 4, 5, 11–15, 23, 27, 28, 32, 36, 39, 45, 46, 48, 50, 54, 60, 66, 71, 75–77, 80–84, 87, 90, 93, 95, 96

M

Mass transfer, 6, 10, 12, 18, 23–25, 27, 31, 32, 51, 53, 66, 68, 70, 71, 75–79, 83, 84, 92, 95, 96

O

Optical measurement, 51–54

P

Parameter variation, 11, 15, 33, 34, 38–40, 69, 75, 95
Permeation, 5, 27–30, 38, 39, 73, 90–92

S

Sublimation, 2, 23, 25–27, 33, 69–71, 75, 76, 83, 87, 95, 96
System effects of defrost, 17–18

V

Vapor transport, 16, 26, 51, 53, 75

© Springer International Publishing Switzerland 2015
W.F. Mohs, F.A. Kulacki, *Heat and Mass Transfer in the Melting of Frost*,
SpringerBriefs in Applied Sciences and Technology,
DOI 10.1007/978-3-319-20508-3

Printed in the United States
By Bookmasters